Ores of the Leadville Mining District

U.S. Department of Interior

Compiled by the staff of
the United States Geological Survey

with an introduction by Kerby Jackson

This work contains material that was originally published by
the United States Geological Survey in 1926.

This publication was created and published for the public benefit,
utilizing public funding and is within the Public Domain.

This edition is reprinted for educational purposes
and in accordance with all applicable Federal Laws.

Introduction Copyright 2014 by Kerby Jackson

Introduction

It has been ninety years since the Department of Interior released it's important publication "Guide to Ores in The Leadville District, Colorado". First released in 1926, this important volume has now been out of print and has been unavailable to the mining community since those days, with the exception of expensive original collector's copies and poorly produced digital editions.

It has often been said that "*gold is where you find it*", but even beginning prospectors understand that their chances for finding something of value in the earth or in the streams of the Golden West are dramatically increased by going back to those places where gold and other minerals were once mined by our forerunners. Despite this, much of the contemporary information on local mining history that is currently available is mostly a result of mere local folklore and persistent rumors of major strikes, the details and facts of which, have long been distorted. Long gone are the old timers and with them, the days of first hand knowledge of the mines of the area and how they operated. Also long gone are most of their notes, their assay reports, their mine maps and personal scrapbooks, along with most of the surveys and reports that were performed for them by private and government geologists. Even published books such as this one are often retired to the local landfill or backyard burn pile by the descendents of those old timers and disappear at an alarming rate. Despite the fact that we live in the so-called "Information Age" where information is supposedly only the push of a button on a keyboard away, true insight into mining properties remains illusive and hard to come by, even to those of us who seek out this sort of information as if our lives depend upon it. Without this type of information readily available to the average independent miner, there is little hope that our metal mining industry will ever recover.

This important volume and others like it, are being presented in their entirety again, in the hope that the average prospector will no longer stumble through the overgrown hills and the tailing strewn creeks without being well informed enough to have a chance to succeed at his ventures.

Kerby Jackson
Josephine County, Oregon
June 2014

GUIDES TO ORE IN THE LEADVILLE DISTRICT, COLORADO

By G. F. Loughlin

INTRODUCTION

The new report on the geology and ore deposits of the Leadville mining district, Colorado, recently completed, is similar in scope to other exhaustive reports on mining districts issued by the United States Geological Survey. The data of most value to mine operators and their engineers, who are primarily interested in the search for ore, are necessarily presented here and there in chapters on stratigraphy, igneous rocks, and structure, as well as in the chapters devoted to ores and ore deposits. In order, therefore, to focus attention more sharply on the problems of ore hunting the present paper is issued, largely at the suggestion of Mr. Augustus Locke, a mining geologist who has been giving much time and energy to "ore-hunting geology" during the last few years.

As operators in the district are thoroughly familiar with the local geology as presented in Emmons's famous monograph and atlas,[1] no geologic map of the district is included with this paper. Certain illustrations, however, of particular interest to mine operators are presented here in advance of the complete report, also the chapter on ore reserves. The general geology is briefly summarized, as the vast amount of data made available since Emmons's report was issued have made necessary certain revisions in the interpretation of geologic structure.

The information on which this paper is based was obtained mainly by the late S. F. Emmons, of the United States Geological Survey, who studied the district in detail at frequent intervals between 1880 and 1910; by the late J. D. Irving, who assisted Emmons in 1901 and later years and prepared the bulk of the new report; by J. E. Spurr, who also assisted Emmons for a comparatively brief period and made independent examinations later; and by the present author, who studied the oxidized zinc ores and fea-

---

[1] Emmons, S. F., The geology and mining industry of Leadville, Lake County, Colo.: U. S. Geol. Survey Mon. 12, 1886.

tures of other oxidized and sulphide ores exposed subsequent to field studies by those already named.

The mining companies and mine surveyors have all been most cordial and generous in supplying data. Mine officials who have published noteworthy contributions to the local geology of ore deposits are F. T. Freeland, A. A. Blow, Max Boehmer, C. J. Moore, Philip Argall, and George Argall. The many excellent mine maps by Howard Platt and J. M. Kleff have been of great aid throughout the district, as have those by F. A. Aicher, formerly of the Iron-Silver Mining Co., in the Iron Hill area.

There has been a striking difference between the work of the Survey geologists and the local engineers named in the preceding paragraphs. Although the geologists' visits have extended over a period of more than 40 years, the aggregate time spent in the district by them has been comparatively very short, and their work has been interrupted by other duties; furthermore, the finding of guides to ore was not paramount but one of several coordinate purposes. The local engineers have had the benefit of more continuous residence in the district, but their observations have generally been confined within small areas. The search for ore has been paramount with them, but it has not always been supported by an adequate appreciation of available geologic data that are important guides to ore. Cordial cooperation between these engineers and the Survey geologists has been mutually beneficial, but the geologists have never had access to much ground that was worked between their visits. The work done, however, forms an adequate basis for a general discussion and appraisal of the guides to ore and for future intensive work by mine operators.

METHODS OF ORE HUNTING USED IN THE PAST

Placer mining began at Leadville in 1860, and some mining of gold veins in porphyry began soon afterward; but the first outcrop of replacement ore in limestone was not discovered until 1874, when lead carbonate was recognized on Rock Hill. A little later similar outcrops of ore were found on Iron, Carbonate, Breece, and Little Ellen Hills, and development proved that they all replaced Blue limestone beneath a sheet of porphyry. (See pl. 1.) The ore at first was followed wherever it led the way. As the porphyry contact undulated considerably, however, and as the ore shoots in places plunged away from the contact or branched irregularly, this method of closely following the ore became awkward; but the porphyry contact at the top of the Blue limestone had now become recognized as the horizon of the ore, and less awkward ways were devised for reaching it. Straight inclines were driven parallel to the eastward dip of the limestone and from them the contact was located by raises or winzes and developed by drifts along the strike. As the inclines became too long for rapid and economical operation, vertical shafts were sunk to reach the ore farther down the dip than the inclines extended. These operations proved the continuity of the Blue limestone eastward beneath an increasing thickness of porphyry and doubtless prompted the sinking of other shafts with the hope that the Blue limestone would be found productive wherever it was reached, particularly near ground that was being mined. Inclines also were driven at different places along the contact, doubtless with similar hope. The contact was the only recognized guide and was soon found to be far from infallible. Localization of ore along the contact was not understood. Ore shoots were accompanied by mixtures of clay and oxides of iron and manganese, known as "contact matter" or "vein material," and the relatively widespread distribution of this material led prospectors to expect ore near by. "Contact matter," if correctly interpreted, would have been of some use as a guide to oxidized ore, but its mere presence was not a reliable guide

Ore was so plentiful, however, that some of these half-blind attempts at ore hunting were successful. In fact, the extensive ore bodies in the Fryer Hill area owe their discovery to totally blind prospecting. They reached the bedrock surface but were covered by glacial débris that averaged 100 feet in thickness. Emmons[1] relates that two intoxicated prospectors sank a shaft haphazard in this débris and discovered ore beneath it at a depth of 25 or 30 feet. Prospectors then swarmed over the area, and ore was found at several places beneath the débris.

Just how much was known of the geologic structure and stratigraphy of the district before Emmons's arrival at Leadville in 1879 can not be stated, but the first systematic geologic description of the district was that given in Emmons's preliminary report in 1882.[2] This report and the more comprehensive monograph issued in 1886 gave so clear and accurate a picture of the geologic structure that to this day the monograph is frequently referred to as the Leadville miner's bible.

Plans for further mine development along porphyry contacts could now be more intelligently made, as could the search for ore that had been cut off by large postmineral faults. The amount of necessary guesswork was greatly reduced, and expenditures for exploration became more efficient. For example, where the "main ore shoot" in Iron Hill ended abruptly against the westward-dipping Iron fault, the McKeon incline was sunk along the fault, and crosscuts were driven westward at successively lower levels. The

---

[1] Emmons, S. F., Geology and mining industry of Leadville, Colo.: U. S. Geol. Survey Mon. 12, p. 13, 1886.

[2] Emmons, S. F., Geology and mining industry of Leadville, Lake County, Colo.: U. S. Geol. Survey Second Annual Rept., pp. 210–290, 1882.

ore body was found to be broken by seven parallel auxiliary steplike faults, and its westward continuation, when once located, was worked through the Satellite shafts. Where the old Mikado ore body was cut off by the Iron fault near Stray Horse Gulch, a shaft was sunk considerably to the west but passed from a hanging wall of White porphyry into a footwall of pre-Cambrian granite and missed the ore-bearing contacts. Another shaft was therefore sunk still farther

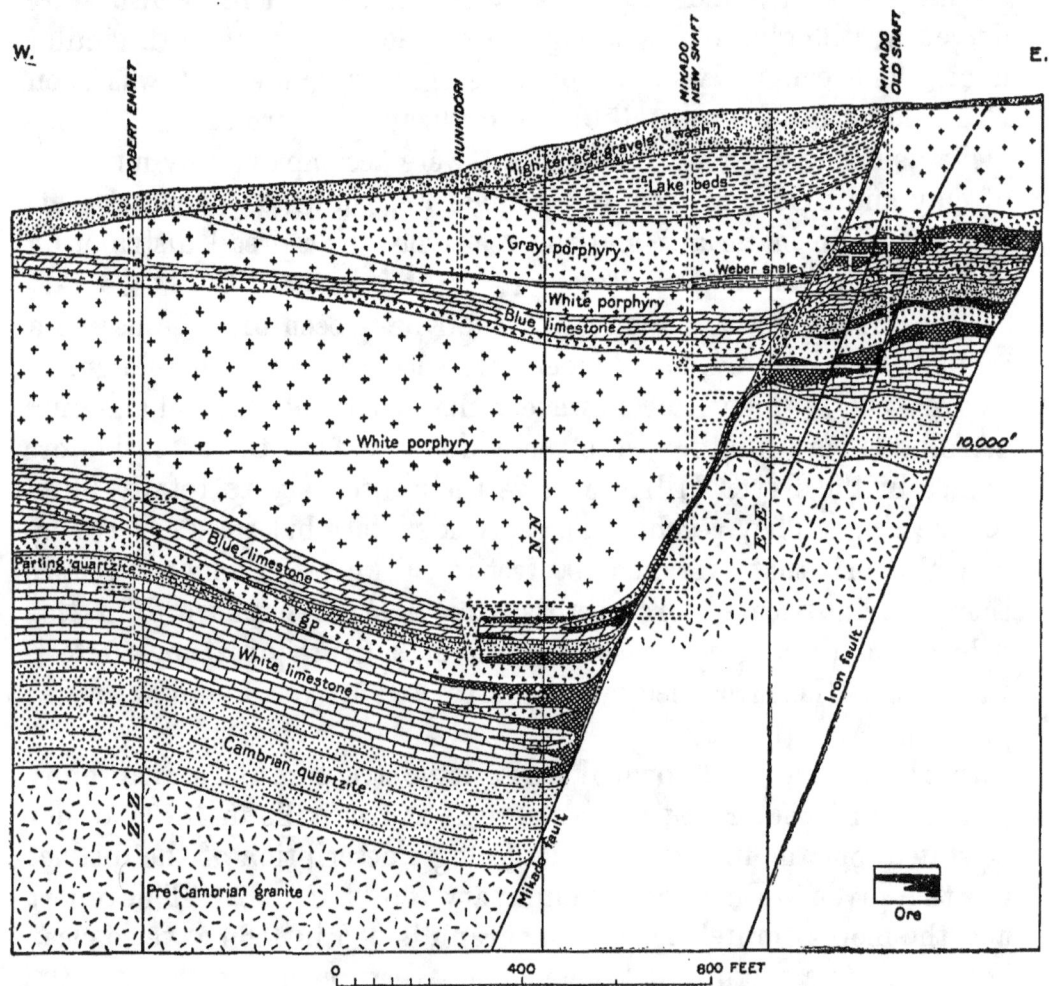

FIGURE 1.—Section through Mikado shafts

west. It cut the hitherto unknown Mikado fault, nearly parallel to the Iron fault, and also passed into pre-Cambrian granite but disclosed enough dragged ore in the fault zone to pay for mining. Crosscuts were run to the fault zone at successively lower levels (fig. 1) until the main ore body in Blue limestone was found. The ore body was then developed westward and found to be continuous with ore that had been followed eastward from the outcrop on the west slope of Carbonate Hill.

The following of ore had disclosed the fact, prior to Emmons's arrival, that some shoots extended downward across the dip to a persistent sill of Gray porphyry, beneath which there was a second "contact" or ore horizon. The White limestone had been penetrated at a few places, at some of which it was mineralized, but the top of the Blue limestone was the horizon of by far the most productive deposits in the district, and Emmons attributed this fact to physical conditions that concentrated the ore-forming solutions along the contact between Blue limestone and White porphyry. This statement still holds true, although the interpretation of the source of ore and the courses followed by ore-forming solutions has changed considerably.

It is significant that although Emmons's views regarding the genesis of the ores were modified considerably in the course of time, by himself as well as others, his data showing the district habits of the ores, so far as they had been revealed to him, have stood the test of time. These data were expressed largely on maps and sections supplemented by the text of the monograph. It was the clear picture conveyed by the maps and sections that impressed and guided the mine operators.

PRESENT GUIDES TO ORE

During Emmons's first survey of the district mine workings were too few and too shallow to reveal adequately the structural relations of the porphyries and the complexity of faulting. The presence of an intrusive stock in the western part of Breece Hill was not suspected, and the significance of adjacent magnetite-specularite ore with serpentine gangue was not appreciated. Certain reverse faults were misinterpreted as normal faults, but their importance as guides to ore was not revealed until several years later. Emmons had given attention to most of these features during his second survey, but he died before his conclusions regarding them had taken definite form. Irving, who was Emmons's principal assistant and who continued the work, also died before its completion, but the manuscript of the new report as left by him contained descriptions and interpretations of nearly all the features that can serve as present guides to ore.

SUMMARY OF GEOLOGY

A list of the sedimentary formations, mainly as worked out by Emmons but with some revision as to age, is given in the table below. A general idea of the distribution of these formations and their relations to the igneous rocks that are intrusive into them may be gained from the atlas accompanying the Leadville monograph. Revised geologic maps and sections will accompany the new report.

Stratigraphic section of the Leadville district, Colo.

Age	Name	Thickness (feet)
Pennsylvanian	Weber (?) formation:	
	"Weber grits"	940
	"Weber shales"	0–300
Mississippian and Devonian (?)	Leadville or "Blue" limestone	200
Ordovician	Yule limestone:	
	"Parting" quartzite member	10–70
	"White" limestone member	120
Upper Cambrian	Sawatch quartzite ("Lower" quartzite):	
	"Transition shales"	0–40
	Quartzite	50–140
Pre-Cambrian	Granite, gneiss, and schist	

These rocks were invaded in late Cretaceous or early Tertiary time by porphyry of two distinct kinds. The earlier, locally called white porphyry, is equivalent to a muscovite granite or salic granodiorite in composition. It formed an immense intrusive sheet between the "Weber shales" and blue limestone in the western part of the district and several relatively small sheets at lower horizons within the Blue limestone, in the eastern part. The source of the White porphyry has not been determined. Emmons, in the first report, believed it to be at Mount Sherman, but Irving inferred that it was the same conduit through which the Gray porphyry was intruded beneath the west slope of Breece Hill.

The later or Gray porphyry includes four recognized varieties, which differ mainly in texture and the presence or absence of certain phenocrysts. These varieties were doubtless intruded at different times, but their structural relations to one another are obscure, and, so far as their relations to the occurrence of ore are concerned, their differences are of no consequence. The Gray porphyry forms a number of intrusive sheets, some of which are very irregular. The most extensive of these sheets overlies White porphyry in the northwest quarter of the district. The others are mostly in the Blue and White limestones and the Cambrian "transition shales." They decrease in number westward, and only one, in the Blue limestone, extends into the Downtown area, beneath the city of Leadville. They increase in thickness and complexity in the southern Iron Hill and Breece Hill areas and are accompanied by dikelike offshoots of considerable size. In the Fryer Hill area dikes without any known connection with intrusive sheets cut the Blue limestone and White porphyry.

The intrusion of these irregular sheets greatly disturbed the sedimentary rocks, particularly the Blue limestone. Blocks of limestone were locally thrust aside, and local fracturing was abundantly produced. This was the earliest of the four periods of fracturing and faulting that have a bearing on the search for ore. The fractures of this earliest period were not systematically mapped when the workings in Blue limestone were accessible, and therefore their significance can not be clearly interpreted. Their possible influence as local channels for the circulation of ore-forming solutions, however, should be kept in mind during development work in the vicinity of irregular masses of Gray porphyry.

Subsequent to the intrusion of the Gray porphyry the region was subjected to folding and reverse faulting. The principal folds formed in and around Leadville were anticlines of north-northwesterly trend, with gently sloping eastern limbs and steep or even overturned western limbs. The western limbs were broken by reverse faults of moderate to steep northeastward dip. The largest of these reverse faults are the London fault, on the east side of the Mosquito Range, and the Weston fault, south of Iowa Gulch. The Mike fault, south of Printer Boy Hill, is another of considerable size. Within the mapped area of the Leadville district three reverse faults have been recognized—the Colorado Prince fault, along the northeast edge of Breece Hill; the Tucson-Maid fault, exposed in mine workings of Iron and Carbonate Hills, and an unnamed fault a short distance west of the Tucson-Maid fault in southern Iron Hill. After the formation of the reverse faults fissures and small normal faults were formed at right angles to them.

In adjacent parts of central Colorado the period of folding and reverse faulting was followed by the intrusion of stocks and batholiths of monzonitic rocks. According to Crawford[3] the comparatively late intrusions are represented in the Twin Lakes district, to the southwest of Leadville, and at Mount Bross, to the northeast. No comparable intrusions have been recognized in the Leadville district, but the distribution of ore bodies, particularly those containing magnetite, suggests the presence of such a late intrusion in the stock at Breece Hill. Emmons and Irving regarded this stock as contemporaneous with the Gray porphyry sills, and the distribution of the sills indicates that they rose through the Breece Hill conduit; but the fact that few underground exposures were accessible, together with the high degree of alteration of the rocks exposed, may have prevented the recognition of later intrusive rock within the stock.

At the same time as this suspected late stocklike intrusion, or shortly afterward, more minor normal faults were formed, many of them with radial arrangement around the stock. There is no means of distinguishing this group of minor faults sharply from the fissures and faults transverse to the reverse faults, as in places the trends of both are parallel. From the ore hunter's standpoint, however,

---

[3] Crawford, R. D., A contribution to the igneous geology of central Colorado: Am. Jour. Sci., 5th ser., vol. 7, pp. 365–388, 1924.

nothing is to be gained by distinguishing them. It is sufficient to know that both are later than the reverse faults and earlier than the deposition of ore, which followed the stocklike intrusion at Breece Hill and took place at higher temperatures close to the margins of this intrusion than at considerable distances from it.

The whole sequence of events, from the intrusion of White porphyry to and including the deposition of ore, took place in late Cretaceous or early Tertiary time. In late Tertiary time normal faulting took place on a large scale and divided the district into a number of fault blocks. The principal faults formed at this time are the Cloud City and the Pendery with its auxiliary faults, described by Emmons and Irving in their bulletin on the Downtown district;[4] the Mikado-Iron-Dome group and its auxiliary faults, the Adelaide, Moyer, Ulster-Newton, and Emmet; the Mike (normal movement on an old reverse fault); the Pilot; and the Mosquito (southern part). The Weston (northern part) and Ball Mountain faults may belong to this group, but their relations to premineral faults and ore deposition have not been determined.

Subsequently to ore deposition and to at least part of the postmineral faulting, eruptions of rhyolitic agglomerate took place, especially north of Breece Hill and northeast of Iron Hill. Four funnel-shaped pipes of agglomerate have been partly outlined by mine workings and have been found in places to cut off ore bodies. These were not known during the first survey of the district.

Climatic conditions during late Tertiary time were favorable for oxidation and enrichment of ores to considerable depths. In Pleistocene time glaciation took place at two or more stages, scouring the oxidized surfaces in the upper part of Evans Gulch and in gulches beyond the limits of the district and depositing lateral and terminal moraines and outwash gravel, which have covered the bedrock to considerable depths in and beyond the lower part of the gulches.

GUIDES AVAILABLE AND THEIR SIGNIFICANCE

The guides now available in the search for ore are described below, not in order of value but in the most convenient order for discussion. In fact, some of them may be more aptly termed "detour signs" than guides.

BREECE HILL STOCK

The approximate position of the Breece Hill stock is represented by the central part of the "pyritiferous porphyry" in the atlas of the Leadville monograph. The stock itself has not been appreciably

---
[4] U. S. Geol. Survey Bull. 320, pp. 26–30, 1907.

productive. It has been explored by shafts as much as 800 feet deep, which were sunk before the presence and dimensions of the stock were realized and were intended to reach limestone beneath the porphyry; also by the Yak tunnel, which crosses it in a northeastward direction at an average depth of 1,200 feet, where it is walled by pre-Cambrian granite. The only ore mined within the stock has been taken from the upper portions of a few small siliceous pyritic veins enriched by chalcocite and gold. These veins strike north-northeast and are walled by pyritic porphyry. They have been presumably located by following float and trenching. The character of the float and outcrops has not been subjected to systematic study, which might result in the discovery of more veins and perhaps bodies of low-grade rock of sufficient size to justify milling.

At the old Antioch open cut which is in pyritic porphyry close to the southeast margin of the stock, oxidized gold ore was mined from a stockwork, which, according to Irving, was formed in shattered rock at an intersection of fissures trending north-northeast and east-northeast. Whether or not there are similar stockworks or bodies of "brecciated ore" within the Breece Hill stock can not be determined from the meager evidence at hand. The discovery of such bodies, as well as lodes, is dependent upon intensive study of outcrops and accessible mine workings.

In the search for large veins and replacement ore bodies, however, the Breece Hill stock is a negative guide. The larger veins are found in fissures and minor faults that were formed in the surrounding sedimentary rocks and intrusive sheets of porphyry as an after-effect either of folding and reverse faulting or of the intrusion of the stock. Adjustments in the surrounding rocks were greater than within the stock, and the fissures in them were therefore more continuous. The irregular replacement bodies in the vicinity are almost entirely in limestone, but most of those nearest the stock are of little or no commercial value. Some blocks of limestone are inclosed in the stock but have been replaced by silicate minerals and the iron oxides, magnetite and specularite. The Blue limestone in the Penn mine, on the north side of the stock, and in the Ibex mine, on the east side of the stock, has also been replaced by magnetite and specularite. These minerals were deposited at an earlier stage than the pyritic gold-copper ores and the silver-lead-zinc ores, and they only served to destroy the limestone before the valuable ores could reach it. The magnetite-specularite masses have been productive only at the old Breece iron mine (part of the Penn group), where a large body has been quarried in an open cut for flux, and in parts of the Ibex mine, where they have been fissured or shattered and recemented by pyritic gold ore.

Farther from the stock the magnetite-specularite disappears. In the Penn mine it is bordered and underlain by mixed sulphide and siliceous ores or their oxidized equivalents, and other bodies of it may be similarly bordered, especially where they are cut by veins of quartz and pyrite with or without appreciable quantities of other sulphides. These veins may expand into replacement sulphide bodies beyond the limits of the magnetite. The magnetite-specularite masses, in short, are guides in that they show that the miner is too near the intrusive stock, but they are not a guaranty that ore will be found at any place along their outer margins. The surer guide is a sulphide vein trending through the magnetite away from the intrusive stock. As the magnetite-specularite masses were formed at an early stage, they served in some places as impervious caps, beneath which sulphide ore was deposited at a later stage.

Information regarding the western and southern sides of the stock is very meager. Much ground that may have been mineralized on the west side was destroyed by a mass of rhyolitic agglomerate that narrows southwestward and terminates along the Mike fault. Considerable mining was done just west of this mass in the early days, but records of it have been largely lost. Farther south the contact of the stock has not been explored. Such explorations as have been made have disclosed thick sills of Gray porphyry, but there is little information regarding mineralization of the limestone.

There is a smaller stocklike mass at the west end of Printer Boy Hill and another in the vicinity of Adelaide, but there is no evidence that mineralization around them was especially intense. The porphyry mass on the north side of Evans Gulch, east of the Silver Spoon shaft, may also be a stock, but it is nearly all covered by glacial débris, and its structural relations are not definitely known.

OTHER SILICEOUS ROCKS

Ore bodies in other siliceous rocks of the district, including porphyry, "Weber grits," quartzite, and granite, are mostly lodes that consist of one or more parallel veins in fissure or fault zones. These lodes are present chiefly in the eastern part of the district, where the proportion of siliceous rock cut by numerous premineral faults and fissures is greater than elsewhere. Stockworks have been productive in "Weber grits" in the South Ibex mine, on Breece Hill, and in Cambrian quartzite and Gray porphyry in the Cord mine, in the southern part of Iron Hill. A few blanket replacement deposits in "Weber grits" have been worked adjacent to lodes.

Surface indications of these ores are similar to those over the Breece Hill stock—débris of silicified and pyritized rock, considerably leached and whitened by weathering. A considerable area is covered by débris of this kind on the west slope of Ball Mountain and over much of Breece Hill. It has been considerably prospected by pits and shallow shafts, but very few productive veins have been found outside of the Ibex mine and its immediate neighbors. No effort has been made to determine the more favorable places for prospecting by a careful study of the leached débris and scattered outcrops.

Veins.—As many productive veins in this area do not reach the surface, the study of outcrops is only a partial guide. Equally or more important is a knowledge of the trends of the different fissure systems. More than 70 veins, some large and productive and others small but of value as guides, have been found in the Ibex mine. Most of them are in the vicinity of the Ibex Nos. 1, 2, and 3 shafts, in a block of ground bounded on the northeast by the Colorado Prince reverse fault, of northwestward trend (p. 7); on the southeast by the Modoc-Elk fault vein, of west-southwest trend, and the Garbutt fault vein, of south-southwest trend; on the west by the crescent-shaped fault vein designated by Irving "Ibex No. 4," which curves from a north-northeast to a west-northwest trend, and the Weston fault, of north-northwest trend. The shape of the block is shown in Plate 2. Most of the veins within this block trend north-northeast; others trend northeast, east-northeast, and north-northwest. The few productive veins outside of this small area also trend north-northeast.

Another guide, which applies to veins in the "Weber grits," is the distribution of shale beds alternating with those of "grit" or micaceous feldspathic sandstone, which has sometimes been called "quartz porphyry" in the Breece Hill area. The Garbutt vein, for example, narrows locally to a mere streak where shale forms one or both of its walls, and its pay shoots pitch at low angles along the more replaceable beds of "grit."

Although prospecting from the surface has been largely unsuccessful in this area of mineralized siliceous rocks, the distribution of veins that have been productive is so scattered and irregular as to give an unavoidable impression that other veins of commercial grade remain to be discovered. The writer's acquaintance with this part of the district is insufficient to warrant more specific suggestions than have been given in the foregoing paragraphs, but crosscutting from the Sunday, Garbutt, and perhaps other relatively deep shafts to prospect beneath areas that appear to be the most mineralized is worthy of consideration. Chalcocite, the principal indication of enrichment, is present on the lowest levels along the Garbutt and Sunday veins, and any veins in the intervening area may be expected to be similarly enriched. No data are at hand regarding the cost of prospecting in this area, a question that can best be left to local operators. The value of the suggestions here offered obviously depends upon the ratio of the cost of prospecting to the quantity and value of metal produced from mines in the vicinity.

Many of the veins in the Ibex mine have been most productive where they pass from siliceous rock into limestone. If the relative positions of limestone and siliceous rock are known, minor veins only an inch or less in thickness are worth following from the siliceous rock into the limestone, where many of them expand into ore shoots more than sufficient to pay for prospecting and development work.

Stockworks.—As the South Ibex stockwork has been a large producer of siliceous pyritic gold ore, the description of guides to similar ore bodies would be valuable, but it must be confessed that such guides are not yet understood. The South Ibex stockwork was discovered by following stringers westward from the Ibex No. 4 vein in the hope that they would lead to a parallel vein. Instead, they led to a large oval body of broken rock consisting of somewhat rounded fragments of "Weber grits" incrusted and partly replaced by ore. Ore of shipping grade was found in the marginal parts of this body, and the east limit of the ore stoped was about 20 feet from the Ibex No. 4 vein, although the rock between the stope and the vein was shattered and crisscrossed by many veinlets. The central part of the broken rock body was similar in general appearance to the marginal parts but contained ore of only milling grade. The stockworks in quartzite and porphyry in the Cord mine had a similar structure and were formed in shattered rock at the intersection of two fissure zones. The exhausted Antioch stockwork was also formed at the intersection of two fissure zones, but no intersection of pronounced fissure systems has been found at the South Ibex stockwork. This stockwork has lenticular horizontal and vertical cross sections, with the longer axes parallel to the strike and dip of the Ibex No. 4 vein. Augustus Locke[5] has recently suggested that the South Ibex stockwork and certain similar ore bodies may be due to a process of natural stoping during mineralization. Solutions capable of dissolving siliceous rock material rose along fissures, which became enlarged by solution of one or both walls. Where support of the walls was thus removed, pressure was sufficient to spall fragments from them. The fragments became rounded by corrosion in the solution, thus making it possible for spalling to continue until the rising solution was no longer capable

---

[5] Written communications Feb. 25 and May, 1925. Mr. Locke plans to publish a paper on "Mineralization stoping" shortly.

of dissolving the rock material, and the rock fragments became sufficiently wedged together to support the walls. During this closing stage of the stoping process ore deposition began, partly by replacement of rock material and partly by incrusting of the rock fragments.

According to this suggestion stockworks should be found at certain places on mineralized fissures where sulphide solutions during their early stages of circulation were most abundant, but there is at present no way of predicting where such favored places are to be found, except along intersections between fissure systems. Intensive study of fissuring, especially in the eastern part of the district, may disclose places where certain fissures were less tight than elsewhere and were therefore relatively favorable for circulation and consequent spalling or "natural stoping"; but so far as data at hand are concerned the discovery of stockworks or ore-bearing breccia is a matter of chance. An abundance of crisscrossing stringers in the walls of a developed vein may lead to a stockwork but from present experience can not be regarded as a reliable guide.

BLUE AND WHITE LIMESTONES

By far the greatest tonnage of ore mined in the Leadville district has come from the limestones, and much more has come from the Blue limestone than from the White limestone, as shown on Plate 3. This statement applies to the total production of the entire district. In parts of the district, notably Iron Hill and Carbonate Hill, during comparatively recent years the output has come dominantly from ore bodies in the White limestone, which averaged of lower grade than those in the Blue limestone. The large ore bodies that have replaced the limestone are mainly low in silica and valuable for zinc, lead, and silver and in part for their excess of iron and manganese over silica, whereas veins and stockworks in or adjacent to siliceous rocks are mainly siliceous and valuable for gold with or without paying quantities of copper and silver. Where veins crossing beds of limestone between roofs and floors of siliceous rock spread into blanket replacement bodies of considerable size the replacement bodies close to the vein fissure consist of siliceous gold ore, which grades within a short distance into zinc-lead sulphide ore, or, in the oxidized zone, into lead carbonate ore. The "gold ore shoot" that formed one of the extensive replacement bodies in Blue limestone at Iron Hill is an exception to this general statement, and its inaccessibility for the last several years precludes a study of it in the light of recent geologic data. Its position and details of outline, however, suggest that its relatively high gold content may have been due to a more direct connection with the Tucson-Maid reverse fault, along which local bodies of siliceous gold-copper ore would be expected.

The difference in productivity of the two limestones is due in minor degree to differences in composition, texture, and structure, but mainly to differences in larger physical features, such as the thickness and continuity of impervious roofs and floors and the number and continuity of premineral faults and fissures.

Differences in composition, texture, and structure.—The Blue limestone is rather thick bedded and is shown by several chemical analyses to be a dolomite which contains little or no free calcite and less than 3 per cent of insoluble matter, whereas the White limestone consists of thin beds of dolomitic limestone which contain considerable free calcite and 10 to 20 per cent of insoluble matter and are separated by thinner beds and partings of shale.

The "transition shales" of Cambrian age, which underlie the White limestone and are regarded by local operators as a part of the White limestone, also contain thin dolomitic beds alternating with a greater amount of shale than is present in the White limestone. These differences imply that the Blue limestone is more subject to complete replacement than the White limestone and "transition shales," and they may account for the statement that ore bodies that replace Blue limestone are commonly of higher grade than the others; but the distribution of ore bodies is due to larger structural features, the principal of which are roofs and floors of impervious siliceous rock.

Roofs and floors.—Replacement bodies have been found at different horizons which differ in number in different parts of the district, as shown in Plate 4. At some places ore has been found only at the top of the Blue limestone, and beds at lower horizons have either been found unprofitable or barren or remain unexplored; at other places ore has been mined at as many as 11 horizons, most of which are in limestone, though here and there small bodies in the quartzites have been mined. Examination of Plate 4 will show that nearly all the ore bodies in limestone are covered by roofs of impervious siliceous rock, most of which is porphyry and the rest shale or quartzite. Some of the thickest ore bodies have floors as well as roofs of impervious rock. A few ore bodies rest on a floor of impervious siliceous rock and pass upward into limestone, and a few others are bounded both above and below by limestone.

These barriers have been of prime importance in localizing the deposition of ore by deflecting and impeding the progress of ore-forming solutions and giving them correspondingly great opportunity to react with the limestone, but the presence of a barrier is not alone a sure guide to ore. The solutions originated in a deep-seated source and moved along a general upward course, but they were locally guided by available openings in the rocks. For the most part the solutions rose to the top of the Blue limestone ("first contact") and were deflected along its top by White porphyry or "Weber shales," but in places they reached fractures that afforded more ready progress downward through the limestone until the solutions were deflected and impeded by a floor of Gray porphyry or "Parting" quartzite. In some places solutions may have found a passage upward through a Gray porphyry sheet or one of the quartzites, only to be impeded and deflected by impervious shaly beds a few feet above in the limestone. These conditions also resulted in an ore body with a floor of siliceous rock and a roof so inconspicuous in comparison that the ore may have appeared to have graded into limestone. No ore bodies wholly surrounded by limestone have been seen by the writer, and, as no thorough descriptions of their boundaries have been found, their positions can not be adequately explained. Careful observations along the boundaries of ore shoots may disclose leads to other ore shoots along the same or another barrier; but to judge from general experience in the past leads from one barrier to another may be so inconspicuous as to be overlooked, and it is therefore worth while to consider whether exploration of barriers above or below a known ore shoot is justified. In general the "first contact" is the most promising; but the others should not be too much slighted, especially in the areas of most intense mineralization, which will be considered presently. The value of a thorough knowledge of the detailed stratigraphy and of local fracturing as a basis for exploration of this kind is obvious.

ORE CHANNELS; TREND AND DISTRIBUTION OF ORE SHOOTS

As barriers or "contacts" could have been effective only where they were reached by ore-forming solution, one of the most fundamental problems in hunting ore in the limestone is to locate the trunk channels and branch channels along which the solutions moved. In the eastern part of the district these channels were obviously the more open parts of normal premineral faults and fissures, which have already been considered, but in the western part of the district many of the channels were in or close by reverse faults.

Channels in the vicinity of the Tucson-Maid reverse fault.—Spurr, in 1907, was evidently the first to recognize the significance of a reverse fault as an ore channel in the Maid of Erin mine, at Carbonate Hill.[6] In 1910 a reverse fault was found in the Tucson

---

[6] Spurr, J. E., unpublished report, 1907; Ore deposition and faulting: Econ. Geology, vol. 11, pp. 601–622, 1916; The ore magmas, vol. 1, p. 353, 1924.

mine, at Iron Hill, by George O. Argall,[7] of the Iron-Silver Mining Co. It was followed downward to the tenth level and upward above the fourth level, where its diminishing dip coincided with that of the Blue limestone. In 1919, Philip Argall and the writer found a reverse fault on the eighth level of the Wolftone mine. It was soon afterward mapped in detail by Frank Aicher, engineer-geologist of the Iron-Silver Co., and found not only to lie in line with the fault found by Spurr but to be the offset northwestward continuation of the fault in the Tucson mine, beyond the post-mineral Iron fault and its nearly parallel auxiliary faults. (See pl. 5.)

As one part of this fault had been called the Maid fault by Spurr, and another part called the Tucson fault by Argall, the entire fault is here called the Tucson-Maid fault. It has been traced southeastward across the Yak tunnel to the workings of the Cord mine, where it appears to be branching and dying out. It has not been traced northwest of the Maid of Erin mine, but the amount of its throw there and local structural details recorded in geologic cross sections imply that its faulted continuation is present in the Downtown area. There is no evidence, however, that it extends as far northwestward as Poverty Flat. In short, this reverse fault traverses the most productive parts of Iron Hill, Carbonate Hill, and the Downtown area.

Examination of the fault, particularly in the Tucson mine, shows that most of it is too full of siliceous gouge to permit ready circulation of waters, but the more open auxiliary fractures along it have served as channels. These fractures are present in all the rock formations and have been especially favorable to the deposition of zinc-lead-silver ore in the White limestone. (See fig. 2 and pl. 6.) Where ore was deposited locally within the Tucson fault itself siliceous pyritic ore formed in contact with the siliceous fault material and graded into zinc-lead-silver ore in the adjacent White limestone. The siliceous ore is enriched by chalcocite, silver, and gold. Fissures and an associated shattered bed of Cambrian quartzite near the fault contained an unusually rich ore in which the common sulphide minerals were accompanied by an intergrowth of silver and bismuth sulphides. The ore contained as much as 14 ounces of gold to the ton.

In the workings of the Cord mine below the Yak tunnel level most of the ore bodies replace masses of White limestone between sheets of porphyry at the intersection of the Tucson fault zone and a mineralized fissure of northeastward trend known as the Cord fault or Cord vein. (See figs. 3 and 4.) The replacement bodies.

---

[7] Argall, G. O., Recent developments on Iron Hill, Leadville, Colo.: Eng. and Min. Jour., vol. 89, pp. 261–266, 1910.

were found by following the Cord vein and were enlargements of that vein, but the productive part of the vein was within the zone of reverse faulting that it crossed. Within the Cambrian quartzite

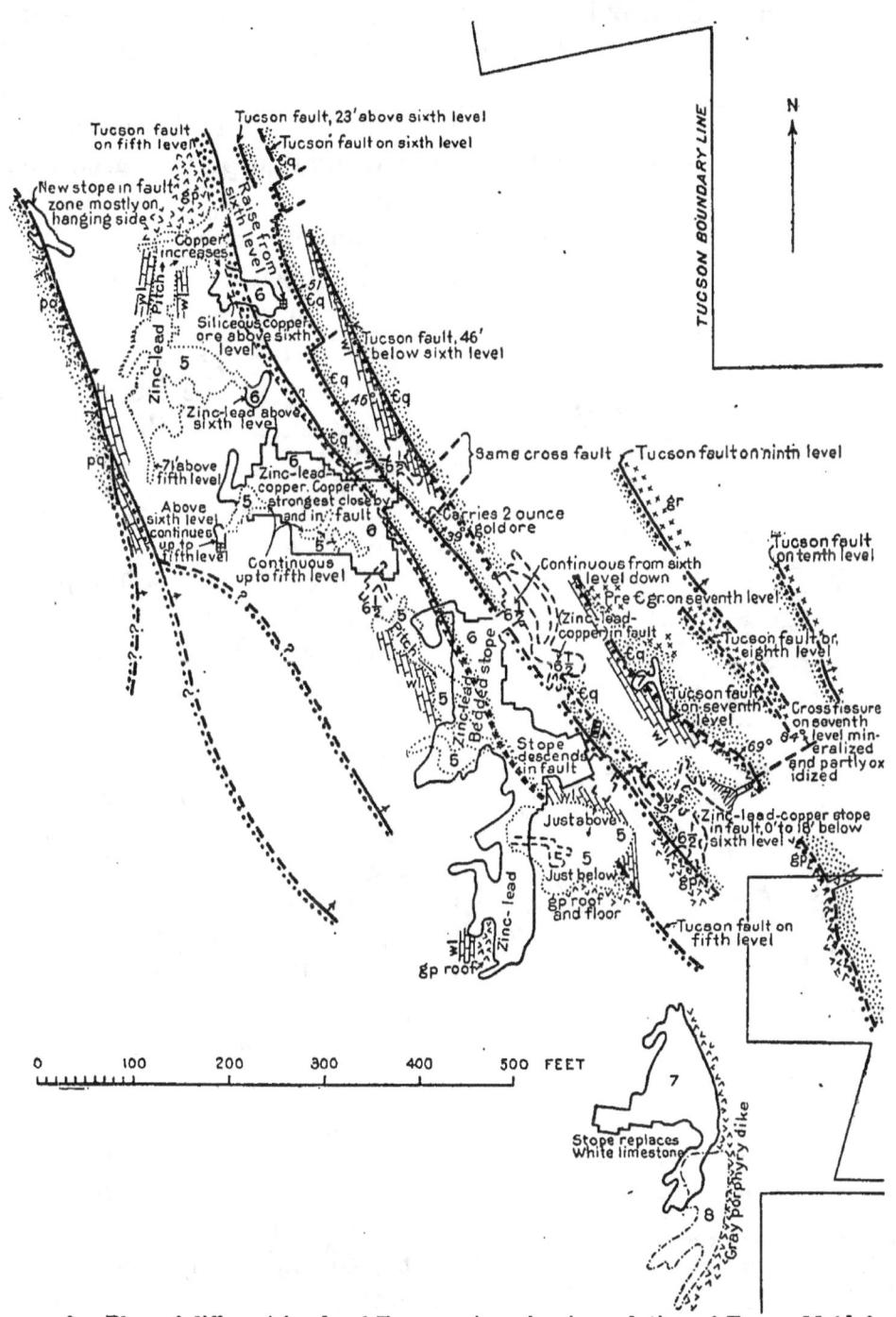

FIGURE 2.—Plan of different levels of Tucson mine, showing relation of Tucson-Maid fault to ore bodies. By F. A. Aicher, Iron-Silver Mining Co. Numbers in stopes refer to levels. pq, Parting quartzite; wl, White limestone; Єq, Cambrian quartzite; gp, Gray porphyry; Pre-Є gr, pre-Cambrian granite

the Cord vein was workable, and between its intersections with two strong members of the Tucson-Maid fault it expanded into a stockwork.

The Cord vein lies directly below one of the long northeastward-trending ore bodies at the top of the Blue limestone in the Iron Hill area and is sufficient evidence that these long ore bodies lie along fissures at about right angles to the Tucson-Maid reverse fault. The evidence in the Tucson and Cord mines indicates that although the reverse fault itself was for the most part too tight to permit circulation of ore-forming solutions, the ground near by was considerably opened. The ore bodies found in these two mines are between fissures of the reverse fault zone on the footwall side. Their absence on the hanging-wall side in the Tucson mine is due mainly to the fact that only quartzite and porphyry are present. At other places where limestone forms the hanging wall ore bodies should be looked for, although the gouge in the reverse fault may have protected the limestone from replacement and have served as a barrier to deflect the solutions into the footwall.

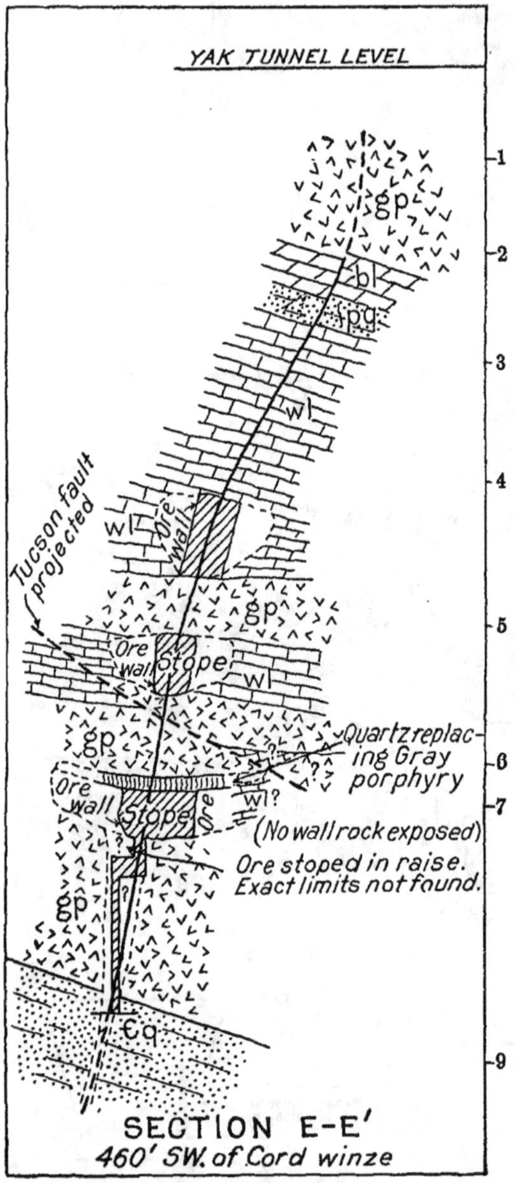

The long ore bodies at the top of the Blue limestone in Iron Hill were mined long ago, and data on the amount of development work done below them are scarce. Detailed geologic sections that will accompany the complete report show the mine workings of which a record could be obtained and imply that at several places in the vicinity of the Tucson-Maid fault the lower part of the Blue limestone and the entire White limestone have not been adequately explored. With the structure shown in Figures 2 and 4 and Plate 6 as a guide further discoveries near the Tucson-Maid fault in the Iron Hill area may be expected.

The great length of the ore bodies at the top of the Blue limestone in Iron Hill and the comparatively small dimensions of the ore

FIGURE 3.—Cross sections of ore bodies along Cord vein. gp, Gray porphyry; bl, Blue limestone; pq, Parting quartzite; wl, White limestone; €q, Cambrian quartzite

bodies in White limestone in the Tucson and Cord mines are due to physical conditions that controlled fissuring. The "transition shales" at the top of the Cambrian quartzite were less subject to fracturing than the typical quartzite itself or the pure dolomitic

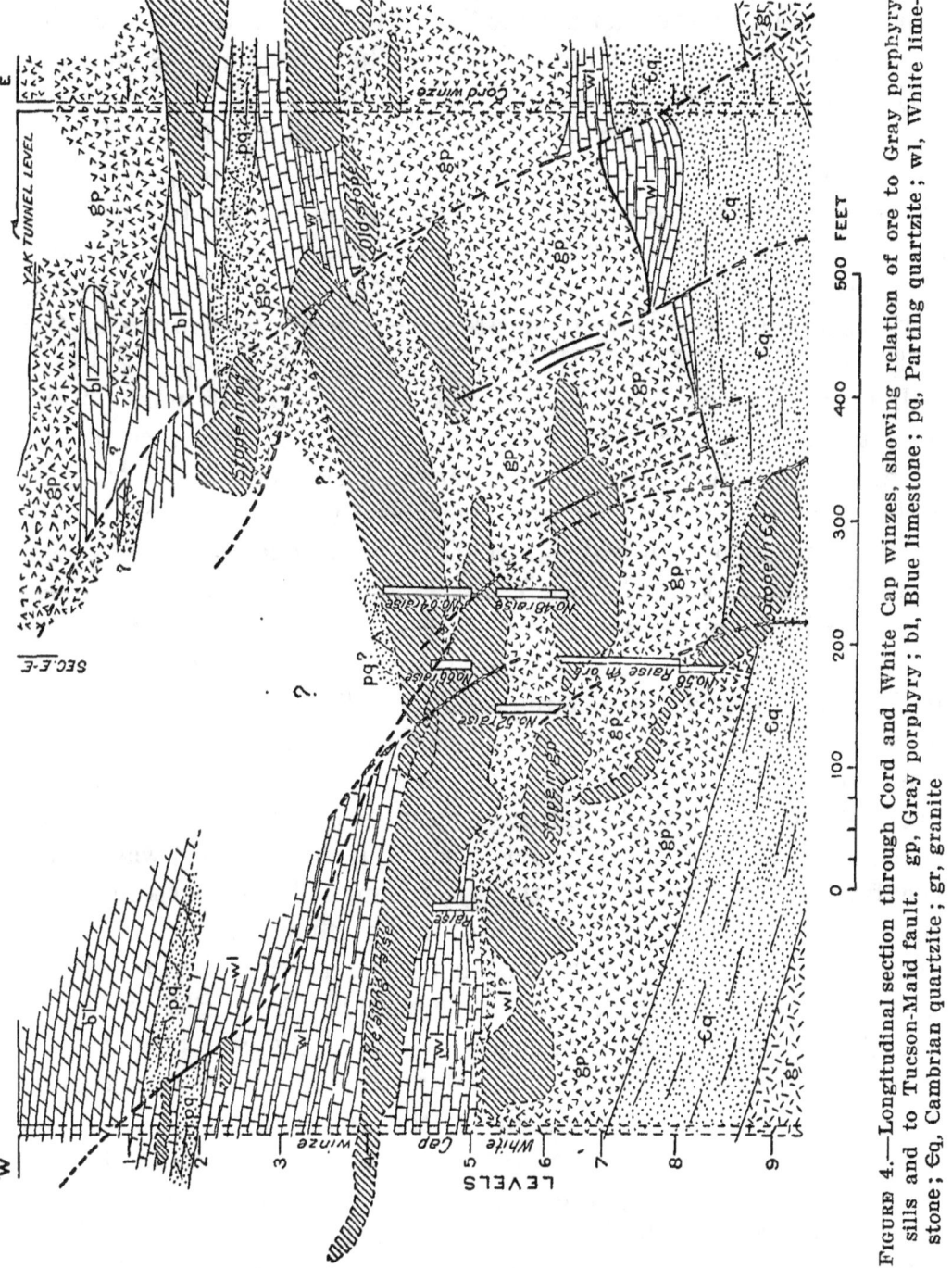

FIGURE 4.—Longitudinal section through Cord and White Cap winzes, showing relation of ore to Gray porphyry sills and to Tucson-Maid fault. gp, Gray porphyry; bl, Blue limestone; pq, Parting quartzite; wl, White limestone; €q, Cambrian quartzite; gr, granite

limestone. The many shale layers alternating with thin limestone beds in the White limestone caused that formation as a whole to resist persistent fissuring to a greater degree than the Blue limestone did. The channels available for ascending solutions in White limestone were therefore most numerous and continuous near the Tucson-Maid fault. The comparatively thin barriers in the White limestone were also considerably fractured in this vicinity, and although they were effective in the formation of ore bodies they allowed a large part of the solutions to rise until deflected by the much more effective barrier at the top of the Blue limestone. The more persistent fissures in the Blue limestone then conducted the solutions for long distances from the places where they had ascended.

Similar conditions existed in Carbonate Hill and are illustrated by Plate 7. Close to the Tucson-Maid fault in the Maid of Erin and Wolftone mines ore was mined continuously from the base of the Gray porphyry sill or "second contact" down to the Cambrian quartzite, but ore did not extend so far from the fault along the lower horizons as along the "first contact," where it extended westward across the Downtown area, eastward to the mines of the Mikado group, in Graham Park, and northward into the Fryer Hill area. The fact that the original sulphide ore is more porous than the country rock implies that deposition of ore, once begun, is increasingly favorable to the continued circulation of solutions so long as no impervious or unreplaceable barriers are encountered.

Development of ore bodies in the Blue limestone at several places in Carbonate, Iron, and Rock Hills shows that some of them taper downward and pinch out along fissures. Emmons noted this feature during the first survey of the district and referred to it as an indication that the ore-forming solutions did not rise along those fissures. In the Rock Hill area the comparatively narrow, veinlike ore shoots in the Blue limestone are southward continuations of the great shoot that extends southwestward from the Moyer mine, in southern Iron Hill. These shoots also pinch out downward, and the few attempts to explore the underlying White limestone have found no encouragement. The ore in the Blue limestone was evidently formed by solutions that had risen in the vicinity of the Tucson-Maid reverse fault to the upper contact of the Blue limestone and then followed along the contact wherever fissures permitted.

Channels in the vicinity of other reverse faults.—The remarkable concentration of ore in the vicinity of the Tucson-Maid reverse fault directs attention to other reverse faults as possible guides to ore. Two other reverse faults are known within the Leadville district—one, unnamed, beneath the southwestern part of Iron Hill, and the other, the Colorado Prince fault, along the northeast edge of Breece Hill.

The unnamed fault beneath the southwestern part of Iron Hill is cut by the Yak tunnel and by the Cord incline about 300 feet west of the Tucson fault zone. Its outcrop is indicated by an offset in the contact between White porphyry and Blue limestone. It has not been explored but is worthy of consideration. If the fault itself or minor fissures near it are found to be mineralized, it is worthy of systematic prospecting, especially in line with the long ore bodies in the Blue limestone. Although the fault has not been explored, the indications that the Tucson-Maid fault dies out southeast of the Yak tunnel imply that the unnamed fault is a steplike continuation of the displacement and is bordered by fractured rock favorable for ore deposition.

The Colorado Prince fault [8] is crossed by the St. Louis vein, short branches of which extend along the fault. At no other places has the fault been found productive, but the fractured limestones along its footwall should be favorable places for ore. The Blue limestone along and near the fault has been mined at the Little Jonny and other mines, but the White limestone has not been adequately explored. The White limestone in the Golden Eagle and Little Vinnie ground, some distance south of the Colorado Prince fault, has been replaced along veins of northward to northeastward trend, and the prospects of finding ore nearer the fault are good.

As all the principal faults of the district were believed by Emmons and Irving to have been formed at one period, they did not look for a continuation of the Colorado Prince northwestward beyond the Weston fault. The bedrock surface west of the Weston fault is all Gray porphyry and is largely covered by glacial débris, so the Colorado Prince fault, if present there, is obscure or concealed at the surface. The domelike anticline with which the fault is associated does not extend much beyond the Weston fault, and the suggested continuation of the Colorado Prince fault is correspondingly uncertain, but some search for it is justified either by drilling or by northward drifts along mineralized fissures in limestones already developed in the Great Hope and neighboring mines.

Besides these reverse faults, the Mike fault, east of Iron Hill, deserves consideration. To the south, in the vicinity of Iowa Gulch, this fault is reverse with northwest strike and northeast dip. North of Printer Boy Hill, where it is joined by the Pilot fault, it is normal but has a downthrow of only 30 feet to the east, although the fault zone is 100 feet wide. The presence of granite boulders in the fault zone between walls of Blue limestone is evidence that granite in one of the walls must have risen at one time to the present level of the Blue limestone. It is therefore inferred that the Mike fault was originally reverse throughout its length but that subsequent (postmineral) settling of the block between the Pilot and Mike faults has destroyed the reverse character of the Mike fault east of Iron Hill. If this interpretation is correct, structural conditions favor-

---

[8] The position of the Colorado Prince fault is shown in the Leadville atlas, but the fault is incorrectly represented as normal.

able for ore may be expected similar to those along the Tucson-Maid fault. Ore has been found in the Mike and Habendum mines near the Mike fault, but the writer has not had an opportunity to study the mine workings in this vicinity and must leave the suggestion to be considered by those acquainted with the ground.

The Weston fault south of Iowa Gulch is a large reverse fault, and veins in it and in auxiliary fissures near it have been worked. The London fault, on the east slope of the Mosquito Range, is another large reverse fault whose auxiliary fissures or faults contain workable veins. Both of these faults are rather remote from the area that has been mapped in detail during the second survey of the Leadville district, but they add to the evidence that the fissures that were produced during and shortly after reverse faulting were favorable for the circulation of ore-forming solutions.

Stringers or "feeders."—Stringers or narrow veinlets of ore have been found at different places in the western part of the district. Irving gave considerable attention to these veinlets during the second survey with the hope that they would prove to be "feeders," filling fissures along which solutions arose to the different contacts; but the absence of such stringers beneath many of the large replacement bodies and the short length of those that had been followed showed that they did not represent the trunk channels of circulation. Some have been found to connect with replacement bodies but to be offshoots rather than feeders. Nevertheless, they are of value as local guides to ore, just as more persistent veinlets in the Breece Hill area have been. Most of them are present in porphyry and quartzite and should be followed upward or downward to unexplored places in the limestones. Although some of these stringers have doubtless been so followed, it appears from Irving's records that others have been neglected. They are worthy of more systematic study than they have received, and they should be followed with due recognition of the kind of rock which they cut. A stringer crosscutting a porphyry sill may lead directly from one replacement body to another. A stringer crossing the "Parting" quartzite may be deflected for indefinite distances along shale layers, particularly near the bottom and top of the quartzite.

At several places where the Blue limestone beneath White porphyry has considerably less than its normal thickness of 200 feet, blocks of it have been detached and inclosed by the White porphyry. At such places stringers in the White porphyry are worth following upward with the hope that they will lead to ore that has replaced the inclosed blocks of limestone. In the Moyer mine a large ore body was found replacing a mass of Blue limestone that had been split from the main mass and raised above the supposed "first contact." This ore body was connected by a stringer with a replacement body in the main mass of Blue limestone. Study of geologic sections on file in the offices of different companies will doubtless show where the main mass of Blue limestone is thinner than normal. A few such places are indicated in Plate 4, and others will be shown in the geologic sections accompanying the complete Leadville report.

SHATTERED ROCK ABOVE REPLACEMENT ORE BODIES

Although small to large vugs as well as a relatively high degree of porosity in the original sulphide ore imply that some decrease in net volume occurred during replacement, no evidence has been reported by geologists or engineers that there has been any appreciable contraction in the limestones during ore deposition. Such contraction would be indicated by a sag in the roof of the ore body and by collapse or shattering of the overlying rock. An appreciable sag may not be noticeable, but the overlying rock may have become considerably fractured and filled with veinlets above the ore body and resemble the stockworks considered on a previous page. Although no such occurrence has been reported, it might well be borne in mind as a possibility when considering mineralized fractures in siliceous rocks as guides to ore in underlying limestone.

In the oxidized zone there has been considerable shrinkage of replacement ore bodies, owing to the complete removal of zinc blende and the partial removal of pyrite from the original ore body, as well as partial removal of adjacent limestone or carbonate gangue. Collapse or shattering of the overlying rock is therefore to be expected. Considerable bodies of shattered White porphyry have been reported but have not been correlated with ore bodies, and the significance of such shattered rock as a guide to oxidized ore can not be closely appraised. Further collapse that has resulted from extensive mining has caused considerable sagging of the ground surface, particularly in the Carbonate Hill area, and suggests what may be expected on a smaller scale in undeveloped areas near Leadville, where extensive oxidized ore bodies are concealed beneath shattered or collapsed rock at comparatively shallow depth.

ALTERATION BORDERS OR CASINGS

Altered rocks in the vicinity of sulphide and oxidized ore bodies are in part rather definite and in part very indefinite guides to ore. Those near sulphide bodies include pyritized siliceous rocks, manganosiderite, and silicified limestone or jasperoid; those near oxidized silver-lead bodies include "contact matter" or "vein material" and bodies of iron and manganese oxides, zinc carbonate, and basic ferric sulphates, which are in part of commercial grade.

As shown on page 10, pyritized siliceous rock, principally porphyry and "Weber grits," is of little direct help as a guide, as the impregnation of pyrite extended for considerable distances from the productive lodes. In porphyry sills also it extends too far from the borders of adjacent replacement bodies in limestone to be of direct help. In fact, it is difficult to find a specimen of porphyry in any of the sills that is not appreciably impregnated by pyrite and the closely associated minerals sericite, chlorite, epidote, and calcite. The limits of the areas containing such altered rock coincide approximately with those of areas containing ore bodies, but other guides must be followed to reach the ore bodies themselves.

Manganosiderite, or iron-manganese carbonate, forms casings around sulphide ore bodies in the intensely mineralized areas. (See pl. 6.) It is particularly conspicuous in the Iron Hill, Carbonate Hill, and Downtown areas in the vicinity of the Tucson-Maid reverse fault. Its principal oxidation product, iron-manganese oxide, or "black iron," is present in the oxidized zones of these areas and at Fryer Hill and elsewhere. The manganosiderite casings range from a few feet to a few yards in thickness. Close to the ore bodies they consist mostly of iron-manganese carbonate, with a small percentage of magnesium carbonate but almost no lime. Away from the ore bodies they pass gradually into Blue or White limestone. They pass abruptly into some ore bodies and gradually into others and surround low-grade pyrite bodies as well as mixed sulphide bodies. The manganosiderite is light gray with a very slight pinkish tinge and is easily distinguished from Blue limestone but is not so readily distinguished from White limestone unless care is taken to note its relatively high specific gravity.

So far as generalization is warranted, it appears that manganosiderite tends to incase ore bodies relatively near the trunk channels along which ore-forming solutions rose, whereas silicified limestone or jasperoid is the prevailing casing at a greater distance from the trunk channels. The inaccessibility of so many of the old mine workings prevents a thorough check on this statement.

The jasperoid casings are less definite guides than those of manganosiderite. In some places the jasperoid forms thin margins of low-grade material bordering relatively large ore shoots; in others the jasperoid forms extensive masses within which small ore shoots are scattered along premineral fissures and are difficult to find.

The "vein material" or "contact matter" found along porphyry-limestone contacts in the oxidized zone is somewhat analogous to the pyritized siliceous rock in the unoxidized zone. It denotes the area within which oxidized ore bodies may be expected, but it may not lead directly to them. The "contact matter" has been largely derived by the leaching action of surface waters descending through the pyritized porphyry and has replaced the limestone. It consists of varying mixtures of iron oxides, clay or "talc" alunite, and jarosite. The same waters that brought in these materials oxidized and leached the ore bodies, staining jasperoid casings brown or black and oxidizing manganosiderite and pyrite to "black iron" and "brown iron" respectively. These materials can not be sharply distinguished from "contact matter."

Where waters after descending through pyritized porphyry reached limestone at a considerable distance from ore bodies they deposited "contact matter," which therefore can not be a direct guide to ore. If premineral fissures can be recognized in limestone containing this "contact matter" they are well worth following. No criteria for distinguishing premineral fissures in the oxidized zone have been established by those who studied the oxidized ore bodies in detail, but their general trends doubtless coincide with the longer dimensions of neighboring ore bodies. Without their aid there is nothing to indicate the direction to search for ore along bodies of "contact matter."

Where the "contact matter" is high in iron and manganese oxides and low in clay or "talc" it is likely to be near an ore body, but the exact position of the ore body may not be evident. The common mode of occurrence of oxidized ore bodies, according to Emmons's descriptions in the Leadville monograph, is for lead carbonate, usually with moderate to high silver content, to lie close to the porphyry roof and to be flanked and floored by "contact matter." Ore bodies with an original jasperoid casing passed outward and downward, after oxidation, into brown-stained jasperoid, below which limestone became replaced to some extent by brown iron oxide or low-grade zinc carbonate. Considerable bodies of pyrite practically free from other sulphides were oxidized to brown iron oxide close to the porphyry roof and surrounded and underlain by brown jasperoid. Shoots of lead carbonate ore are to be looked for inclosed within these brown iron ore bodies along premineral fissures, if such fissures can still be recognized.

Ore bodies originally incased in manganosiderite have less simple relations after oxidation. Pyrite bodies were oxidized to brown iron ore, but at their floors and margins "contact matter" derived from the overlying porphyry and from any gangue in the pyrite replaced the manganosiderite. Any unreplaced manganosiderite was oxidized to "black iron" ore. Where mixed sulphide bodies became oxidized the lead and much of the iron remained in place. A fringe of "contact matter" formed at their floors, and below this fringe zinc carbonate, carried downward from the original ore body, extensively replaced manganosiderite or any limestone within its reach. Any manganosiderite that escaped replacement by zinc carbonate was subject to oxidation into "black iron," which might lie beside or even below zinc carbonate; but as a rule the extensive deposition of zinc carbonate extended well down to the bottom of the oxidized zone or even somewhat below it.

The foregoing brief statement illustrates the complex relation of the oxidized ores, and their intelligent development requires a thorough acquaintance with the variations in structure and composition of sulphide ore bodies and their casings. In general, however, the oxidized ore shoots in limestone may be placed in two classes—those with casings or floors of brown-stained jasperoid and those without them. Where the jasperoid casings are present, lead carbonate and brown iron ore, with high or low silver content, are confined within or above the casing, and fractures in the jasperoid may contain some silver in commercial quantity. Low-grade zinc carbonate will be found here and there below the casing, and its tonnage ratio to lead carbonate ore will be low compared with the ratio of zinc to lead in the average sulphide ore body. Where manganosiderite as well as jasperoid accompanied the original ore body "black iron" with high silica content will be found. Oxidized ore bodies derived from mixed sulphides without jasperoid casings include lead carbonate underlain by zinc carbonate and bordered in places by brown iron ore and "black iron" or iron-manganese ore relatively low in silica. Those derived from practically pure pyrite consist of brown iron ore underlain and bordered by black iron-manganese ore.

Although most of the lead was oxidized to carbonate without being transferred, insignificant quantities were dissolved, carried downward, and redeposited in small part as small veins, bunches, and isolated crystals of cerusite and anglesite. A considerable though minor quantity was deposited as the lead-iron sulphate, plumbojarosite, directly below the lead carbonate bodies, and this was underlain in turn by iron and manganese oxides. The plumbojarosite occurs in earthy yellow masses and is mixed with varying amounts of similar appearing minerals of the jarosite group. In places the percentages of lead and silver are high enough to justify mining.[9] Very little of this material has been seen during the second survey of the district.

ENRICHMENT IN OXIDIZED AND SULPHIDE ZONES

Enrichment in the oxidized zone is largely residual but is in part due to downward concentration. Most of the lead carbonate ore bodies worked in the earliest years were richer in silver than those worked later. The silver was largely converted to chloride and

---

[9] Ricketts, L. D., Ores of Leadville, pp. 36–37, pl. 1, Princeton, 1883.

bromide without being removed from the lead carbonate shoots, but the remarkably high silver content in those bodies nearest the surface suggests that silver was in part concentrated downward as the outcrops of ore bodies were lowered by erosion. After mining had penetrated considerable distances along the dips of the ore bodies the silver content of the lead carbonate ore was appreciably less. Small amounts of silver chloride and bromide were transported short distances and redeposited in cracks in brown jasperoid.

In the transition zone between the oxidized and sulphide blanket ore bodies some native silver and argentite were concentrated, but below the transition zone the sulphide ore in the blanket bodies was not appreciably enriched. In the vicinity of the blanket ore bodies, both in the Tucson mine, at Iron Hill, and the Henriette-Maid mine, at Carbonate Hill, some very rich ore has been found in the Cambrian quartzite. In the Tucson mine the ore was found in veins and connected small vuggy bodies along a shattered stratum. The high silver content, exceeding 1,000 ounces to the ton, was due to an intergrowth of argentite and bismuthinite, known as "lillianite," which lined the vugs.[10] The ore also contained as much as 14 ounces of gold to the ton, but no visible gold was found. These remarkably high contents of silver and gold suggested enrichment, but nothing visible in the ore could be pointed to as a proof of enrichment.

The rich ore body in Cambrian quartzite in the Henriette-Maid mine, according to Emmons, occurred in veinlike form and contained a little native copper, which was evidently deposited by descending waters. The high grade of the ore, therefore, may well have been due to downward enrichment.

In the Tucson mine along the Tucson-Maid fault (pl. 6) pyritic ore well below the bottom of the oxidized zone, which is at the fourth level, is enriched by chalcocite, silver, and gold—in fact, chalcocite may be considered an indication of high contents of gold and silver. Similar ore has been found at the junction of the Tucson-Maid fault and Cord vein in Cambrian quartzite, but the gold content there is said to be erratic and to reach a maximum of 2 ounces to the ton, both in ore that contains chalcocite and ore that shows no sign of alteration. In the Greenback mine, in Graham Park, a vein of mixed sulphides in a branch of the Tucson fault, directly below a large pyritic replacement body in White limestone, is similarly enriched by chalcocite and silver. These occurrences are similar to those in veins of the Breece Hill area, where chalcocite has been found at the lowest levels (Yak tunnel level), far below the oxidized zone.

---

[10] Argall, G. O., Recent developments on Iron Hill, Leadville, Colo.: Eng. and Min. Jour., vol. 89, pp. 261–266, 1910.

Sulphide enrichment was evidently confined to pyritic ores containing small percentages of chalcopyrite and occurred only in lodes and adjacent parts of connected replacement bodies where a comparatively open structure and marginal impervious clay selvages concentrated downward circulation. Copper leached from the oxidized zone was redeposited as chalcocite, which in turn served as a strong precipitator of any silver and gold in the descending surface waters.

Gold is also found concentrated in bunches of mixed sulphide ore, where it is closely associated with zinc blende. This gold was evidently carried downward from the oxidized zone as soluble chloride and deposited by means of a complex reaction involving ferrous sulphate and zinc blende. The result was the replacement of zinc blende along crystal surfaces and fractures by flakes of gold. The rich bunches of gold found in the oxidized zone were probably concentrated in the zone of sulphide enrichment at an early stage of alteration and remained as residual gold after the bottom of the oxidized zone had migrated below them at a later stage.

As enriched ore extends to the deepest levels of mining in the larger, more open veins, the search for it is largely guided by the factors already noted, which control the dimensions of the veins. The broader parts of the veins, if accessible to downward circulating waters, permitted the greatest amount of enrichment. Narrow parts and mere stringers too small to mine also give high assay returns.

OUTCROPS

During the resurvey of the Leadville district little attention was given to outcrops of the mineral deposits, largely because outcrops in the most developed areas were worked out in the early days and many of them were later covered by mine dumps; also because a large part of the district is covered by thick glacial deposits, and the search for ore beneath them has been conducted either by random shaft sinking or by underground exploration from developed ground. The outcrops of silicified porphyry and "Weber grits" in the Breece Hill area, considered on pages 9–10, are the only ones in the resurveyed area that continue to deserve careful attention in development from the surface.

The scant attention given to outcrops during the resurvey, however, should not obscure their importance as guides to ore in the early days of the district, or in the future prospecting of outlying areas. Knowledge of the different products of primary mineralization and later oxidation forms the basis for a better interpretation than was formerly possible, not only of actual exposures of ore and "contact matter" but of collapsed or shattered areas above concealed ore bodies.

ORE RESERVES[11]

After a mining district has been a large producer for nearly half a century it must be admitted that there has been considerable opportunity to determine the limits of productive territory by mining and prospecting and that probably most of the favorable ground within those limits has been thoroughly explored. There are reasons, however, for hoping that the productive area of the Leadville district may still be enlarged and that some good ground heretofore overlooked may be found within the developed area. The relative promise of ground hitherto neglected or inadequately prospected must be gaged by the geologic study of the developed ground, together with a consideration of the cost of exploration. Besides the reserves of promising ground that are believed to exist and the developed and partly developed ore bodies that are minable with profit at present prices it is reported that large quantities of low-grade material are ready to be mined when higher prices or improvements in metallurgy shall make it profitable to treat them. The following remarks will be confined to promising ground awaiting development and ground shown by development to be discouraging.

The discovery of the large ore shoot that had replaced a limestone inclusion in White porphyry in the Moyer mine suggests that a search for other replaced inclusions would be worth while. They should be expected where the Blue limestone below the White porphyry is abnormally thin. The splitting of the Blue limestone into several slabs inclosed in White porphyry at Fryer Hill, the small aggregate thickness of the slabs, and the general lack of information on the underlying rocks raise the question whether the uppermost part of the Blue limestone has been eroded or whether additional inclusions of it, possibly mineralized, remain to be found. The presence or absence of such inclusions can be determined only by study of geologic sections and mine records, verified by drilling.

A more promising field of search is the prospecting of limestone along reverse faults, particularly in the areas between the Tucson mine and Iron fault in Iron Hill and between the Iron fault and the Wolftone and Greenback mines in Carbonate Hill. Some shafts, including the Experiment and City of Paris, have been sunk in these two areas without penetrating the White porphyry. The Cumberland passed from White porphyry into granite at the Mikado fault and therefore lies between the promising parts of the two areas. The influence of structure on the occurrence of ore in these areas should be similar to that in the mines named. (See pp. 15-20.)

---

[11] This section is the concluding chapter of the complete report.

As the Tucson fault shows signs of dying out southeast of the Cord mine, the opportunity for deposition of ore at its intersection with fissures below the "first contact" and "second contact" shoots may have been less than in the Cord and Tucson mines; nevertheless, if costs of development in the White limestone are not prohibitive, the ground southeast of the Cord mine is worth prospecting, especially below the great Moyer shoot. The reverse fault that appears to parallel the Tucson fault on its west side (p. 20) may hold similar relations to ore bodies. Very little is known of this fault at present, but it may be feasible to follow the fault to its supposed intersection with the northeastward-trending fissures beneath the known ore shoots.

The continuity of the Moyer shoot with ore shoots in the Blue limestone in Rock Hill raises the question of similar continuity of any ores that may be found in the White limestone. Almost no work in the White limestone has been done there, and knowledge of the local structure is deficient, but the fact that the ore shoots mined have been comparatively narrow and have pinched downward suggests that the solutions that deposited them moved southward along the Blue limestone from the vicinity of the Moyer shoot and did not enter the White limestone. The fact that the little work done in the White limestone in the Oro La Plata and Stevens mines was unproductive supports this suggestion.

In the area between the Moyer shoot and the small stock of Gray porphyry that contains the Printer Boy lode the presence of high terrace gravel conceals any evidence of mineralization at the surface, and the deep burial of the Blue limestone has retarded prospecting. This area may be beyond the southeast end of the Tucson fault, but the presence of ore on both sides of it suggests that it also may be mineralized; nevertheless, the structural conditions, though little known, do not appear more favorable here than in the Stevens mine, and prospecting is likely to be expensive and very uncertain of success.

The continuity of the Tucson-Maid fault northwest of the Maid mine has not been proved, although the fault, if it persists so far, should be close to certain of the ore bodies in the Downtown area. If projected farther northwestward, with due allowance for offsets along the Pendery fault zone, it should approach the Delante No. 1 shaft; but explorations through that shaft and the Carleton, Seeley, Neptune, and Villa shafts, near by, have failed to find ore, and it must be inferred that no premineral faults or fissures in this vicinity were of sufficient continuity to serve as ore channels. A drift was driven for nearly 2,000 feet from the Delante No. 1 northward to the Hofer No. 1 shaft, and 14 drill holes were sunk in its vicinity without finding encouraging evidence of mineralization. The only such evidence reported in the vicinity is some low-grade manganese oxide ore carrying 1 to 9 ounces of silver to the ton found in the Jason shaft, and a little low-grade pyrite found in the Sequa shaft and in the Stumpf drill hole, near the Elgin smelter. The whole area is aptly named Poverty Flat, and the limits of appreciable mineralization east and south of it approximately coincide with the boundaries of the Fryer Hill and Carbonate groups of ore bodies indicated in Plate 3.

The limits of minable ground west of the Tucson-Maid fault coincide roughly with the west and southwest limits of the ore bodies of the Carbonate and Iron and Rock Hill groups shown in Plate 3. The western limit of mineralization in the Downtown area has not been closely determined, but all indications point to a gradual decrease in quantity and value of the ores west of the Cloud City fault. Work in this direction has been greatly hindered by excessive amounts of water. The local rocks are very permeable, and the fault has been cut so as to allow water from the east side to flow into the lower ground on the west side. Exploration west of the Cloud City fault has been confined to periods when deep working was in progress east of the fault. Shafts were then sunk to water level, but no attempts were made to go farther, as the cost of pumping for any single enterprise was prohibitive. Drill holes were put down west of the fault by the Western Mining Co. in the more westerly workings of the Coronado, Sixth Street, and Penrose mines, but the quantity and grade of ore found were not very encouraging, and the possibility that the ore-bearing rocks might be eroded within a short distance to the west rendered further development inadvisable.

Toward the south some ore was found in the Valentine mine about 400 feet west of the Cloud City (Valentine) fault, but none in the Home Extension mine near by. The A. V. mine, 800 feet farther south, produced some manganese ore for a short time during the World War. The Maple Street shaft, west of the Valentine, passed through nearly 500 feet of " wash " and " lake beds " and 50 feet of White porphyry, reaching the top of the Blue limestone at a depth of 548 feet. A 5-foot thickness of iron oxide, with very low content of silver or lead, was cut there, but it proved to be discontinuous, and this result, together with the difficulty of handling the water, discouraged further prospecting. In short, the available evidence, though scattered, all points to the dying out of mineralization westward and to the improbability that there are any ore bodies west of the Cloud City fault of sufficiently high grade to justify the high cost of prospecting for them.

Only a few ore bodies have been found in the southern part of Carbonate Hill, and none have been found south of California

Gulch and west of the Iron fault, although these areas have been considerably prospected. In southern Carbonate Hill extensive exploration of the Blue limestone has been conducted through the Toledo Avenue, Modest Girl, and California Gulch shafts without finding any large ore bodies, though a small one was found in the Modest Girl. Another small ore body was found in 1894 in the Thespian mine, but with this exception the mine was devoid of mineralization. To the north a large ore body was found in the Evelyn mine at the second or Gray porphyry contact, and it may have connected with the large ore shoots of the Wolftone and Castle View mines; but no maps of the Evelyn and Castle View mines have been available.

In the Toledo Avenue mine the usual "contact matter" between White porphyry and Blue limestone was found but no ore. The Rose Emmet shaft, south of the Thespian, when examined in 1901 was 780 feet deep and had penetrated the first and second contacts of the Blue limestone, and drifts on the 475-foot level extended along the first contact for 176 feet westward, 190 feet southward, and 415 feet northward, but no ore was found. According to Emmons,[12] the Prospect incline and the Rosebud and Deadbroke tunnels on the north side of California Gulch and the Jordan and Swamp Angel tunnels on the south side followed the "first contact" and exposed "contact matter" but no valuable ore bodies.

The nearest ore to the east is the shoot in the Satellite and Star of the West mines, which is a down-faulted continuation of the North Iron ore shoot. To the south of the ore bodies in the Reindeer and Bessie Wilgus shafts is a westward down-faulted continuation of the shoot worked through the Dome incline and the Rock No. 1 and No. 2 shafts. Between these two down-faulted ore bodies extensive explorations have been conducted west of the Iron fault, through the Little Delaware, Hope, Zulu King, Commercial Drummer No. 2, Switzerland, Hawkes, Moffat, Coon Valley, Ontario, and McKeon shafts. A small amount of ore was found in the Hope ground, just southeast of the Little Delaware, but no other noteworthy discoveries were made. Either the southwestward-trending ore channels or fissures of the Iron Hill group, except the two noted above, died out to the east of this ground, or the solutions deposited their metals before reaching it. To the south the Blue limestone has been largely removed by preglacial erosion.

The great group of ore bodies that spreads from the Tucson-Maid fault zone in Carbonate Hill is known to be connected in some places with the Fryer Hill group, and the old workings beneath Stray Horse Ridge, were they now accessible, would probably supply evi-

---

[12] U. S. Geol. Survey Mon. 12, p. 412, 1886.

dence of other connections. As stated on page 20, ore-forming solutions may have traveled northward along the bedding from Carbonate Hill to Fryer Hill, although there is some reason for suspecting a local deep-seated source of the ore at Fryer and East Fryer Hills. No large mineralized fissures have been found there, however, and the meager amount of exploration in the White limestone has been very disappointing. Any undiscovered local trunk channels were evidently not seriously obstructed by the Parting quartzite, and the solutions reached the Blue limestone before being impounded and forced to deposit their metals. There has been less exploration of the White limestone at Fryer Hill than at East Fryer Hill, and the available data are too few to warrant a definite opinion as to the possible presence of ore. If the dikes in these areas, shown in the atlas of the Leadville monograph, are of deep-seated origin and not offshoots of a sill, the fissures along which they rose may have been reopened and served as ore channels, as suggested by J. E. Spurr;[18] but there are no strong indications that the White limestone contains valuable ore bodies. The conditions at Fryer Hill and East Fryer Hill may be similar to those at Rock Hill (p. 20), where continuous though narrow ore shoots were found in the Blue limestone, but the White limestone contains little or no ore.

North of Fryer Hill, in the vicinity of Little Evans Gulch, there are outcrops of silicified limestone, and it is reported that considerable oxidized silver ore was mined there before the decline in the price of silver in 1893. This area was not studied closely during the resurvey, and it remains for future study to determine the factors controlling the occurrence of this ore and whether or not it is of commercial value below the zone of oxidation. Farther north the Roseville mine, on Canterbury Hill, is said to have produced silver ore in the early days, but exploration of this mine through the Canterbury Hill tunnel in 1924 was a complete disappointment. This tunnel, however, at 2,700 feet from its portal, has exposed a low-grade oxidized zinc-lead deposit with a barite-quartz gangue, which is characteristic of deposits that have been profitably worked in other outlying parts of the district.

East of the Carbonate Hill group there may have been a local trunk channel in the vicinity of the Adelaide mine, but the almost total lack of data regarding mines in the Adelaide group, in spite of the large amount of work done, makes it unsafe to offer any suggestions for further developments. North of the Adelaide group and east of the Fryer Hill group explorations in the Yankee Hill area have been disappointing.

---

[18] Written communications.

The Adelaide and Iron Hill groups are bounded on the east by a stocklike mass of pyritic porphyry, in which small veins may be found, but the amount of unproductive work already done shows that the chances of success in prospecting are relatively small. Only the enriched parts of the veins are likely to be of commercial value, and the zones of oxidation and sulphide enrichment are shallower here than elsewhere in the district. The only limestone inclusions found within this porphyry have been replaced by silicates and magnetite, and the prospects of finding deposits of commercial value in them are small.

East of this stock the much faulted and fissured Ibex area has been intensively but thus far not exhaustively explored. Structural conditions north of the Ibex No. 4 and west of the Little Vinnie shaft are favorable for the presence of several small veins and connected blankets of siliceous ore similar to those mined in the Golden Eagle mine. The junctions of such veins, if present, and the Colorado Prince reverse fault should be favorable places for ore. It has been a curious fact that the Colorado Prince fault has thus far been cut only where its walls are siliceous rocks, and the most favorable ground—limestone at the junctions of the fault with mineralized fissures of northward trend—has not been touched, so far as available records show. The junction of the Blue limestone on the Colorado Prince fault has been entirely removed by erosion, although the blanket ore bodies in the Little Jonny mine are not far from its former position. The junction of the White limestone and the Colorado Prince fault is also removed near the Modoc lode and near the Weston fault, but it is well worth prospecting in the intervening ground.

The possible continuation of the Colorado Prince fault west of the Weston fault, mentioned on page 21, can be determined only by systematic drilling, as structural evidence is concealed by porphyry and glacial débris. If its continuation is established, its junctions with veins cutting the limestones should be favorable locations for ore. The little information available about the Great Hope and neighboring properties on the slope south of Evansville indicates that this area has received less attention than it deserves, largely because of the shallow water level. The fissures through which ore in this area was introduced should continue northward as far as the suggested continuation of the Colorado Prince fault. Although these structural relations are promising, the ore horizons are beneath Evans Gulch, where ground water is doubtless abundant and the zone of oxidation shallow, owing to its partial removal by glacial erosion. The proportion of enriched ore may therefore be relatively small, and it is a question whether or not the grade of the primary ore may be encouraging. The interpretation of concealed geologic structure commends this area for consideration, and it remains for systematic drilling to determine whether further development work is justified. If plans to drive the Canterbury Hill tunnel into this area are realized, the drainage problems may be largely solved.

Further continuation of the Colorado Prince fault would bring it near the Mammoth Placer shaft, which is said to have cut low-grade and also some high-grade ore, a result which encourages the hope that favorable conditions similar to those just considered may exist there also; but a fault of the size and character of the Colorado Prince fault can not persist indefinitely, and not much interest in this vicinity is justified until favorable results have been obtained from prospecting at Evansville.

Another block of ground where structural conditions appear favorable for the presence of ore is that south of the Modoc vein and east of the Garbutt vein, where faulting has carried the blue limestone below the levels that have been productive in the Ibex mine. The position of the limestone can not be accurately calculated, owing to imperfect knowledge of the local structure and to the indefinite thickness of the overlying Weber (?) formation and porphyry sills; presumably, however, it could be located by systematic drilling from the surface or by drifting to the east of the Garbutt vein. This ground is presumably most fissured, and therefore most mineralized, near its northwest corner, just beyond which a number of veins have been found in the Ibex mine. The rocks at the surface here are considerably silicified, and it would not be surprising if ore bodies were found replacing the limestone for some distance east of the Garbutt vein.

The mineralization at the surface continues southward and southeastward beyond the Sunday lode, and it is reasonable to expect other veins between the Garbutt and Sunday veins, but structural details have been so obscured by "wash" that the veins, if present, can be found only by very close study of the débris, followed by trenching. The limestones in the vicinity of the Sunday vein are deeply buried, probably below the level of the Yak tunnel, and prospecting of them would necessarily be expensive.

Northeast of the Colorado Prince fault there may be additional blanket deposits branching from the west as well as from the east side of the Winnie-Luema vein; but prospecting, especially on the west side, is largely a hit-or-miss undertaking. Extensions of the blanket ore bodies in Blue limestone may exist, and other similar bodies may be found between those shown on Plate 3; but here, as elsewhere in the district, ore shoots of which there is no record may have been exhausted in the early days.

The northernmost workings in the Winnie-Luema vein indicate that the limit of ore deposition has been reached and that only low-grade jasperoid is likely to be found farther north. The low grade of the ore in the Diamond mine, together with the increasing depth of the Blue limestone toward the northeast, gives little encouragement for development any farther in that direction.

No systematic study of the territory beyond the limits of the Leadville district has been made during the resurvey. This territory includes a few mines that have been notable producers but are handicapped by lack of transportation facilities. No appraisal of this outlying territory can be made here, but it is hoped that the information presented in this report will serve as a basis for more extended studies by those who are interested.

INDEX

	Page
Alteration casings, nature of	23–26
Breece Hill, intrusion at	7
Breece Hill stock, description of	8–10
ore in and near	9–10
Carbonate Hill, exploration on south part of	32
mineralizing channels in	20
promising area on	29
Chalcocite enrichment indicated by	11
Channels, mineralizing waters follow	15
near the Tucson-Maid reverse fault	15–20
Cloud City fault, exploration west of	31
Colorado Prince fault, possible continuation of	34–35
possibilities of ore near	21
Contact, use of, as a guide	2–3
"Contact matter," nature and occurrence of	24–25
Cord mine, mineralizing channels in	16–20
promising area near	30
Débris, ore beneath	3, 10–11
Emmons, S. F., study and reports by	1, 3, 5
views of, on the genesis and concentration of ores	5
Enrichment, products of	26–28
Faults, production of	7, 8
reverse, promising fields near	29
relations of, to ore bodies	15–22
Floors, ore bodies underlain by	14–15
Folds, formation of	7
Fryer Hill area, discovery of ore in	3
exploration of White limestone in	33
Garbutt vein, possibilities east of	35
prospecting near	11
Geology of the district	5–8
Hunting for ore, past methods of	2–5
Ibex area, possibilities of ore in	34
Iron fault, following of ore bodies cut off by	3–4
Iron Hill, promising area on	29
reverse fault beneath	20–21
Jasperoid casings, occurrence of	24–26
Limestone, Blue, disturbance of	6–7
Limestones, composition of	14
productivity of	13–14
veins entering, from siliceous rock	12
London fault, features of	22
Magnetite-specularite masses, indications from	10

	Page
Magnetite-specularite masses, mining of	9
Manganosiderite casings, occurrence of	24, 26
Mikado fault, following of ore body cut off by	4
Mike fault, description of	21–22
Ore, deposition of	8
Ore bodies, cut-off, methods of following	3–4
stoped, map of	In pocket.
Outcrops, availability of	28
Oxidized zone, enrichment in	26–27
Plan showing the more important veins and faults in the eastern part of the district	10
Plumbojarosite, occurrence of	26
Porphyry, Gray, invasion by	6
White, invasion by	6
Poverty Flat, exploration on	30–31
Reserves of ore, probable extent of	29–36
Rhyolite, eruptions of	8
Roofs, ore bodies overlain by	14–15
Scope of the report	1–2
Sections along the most important inclined shafts	10
columnar, showing position and relative size of blanket ore bodies	16
Shale beds, distribution of	11
"Shales, transition," features of	14
Shattering of rock overlying ore bodies	23
Siliceous rocks, ore bodies in	10–13
Silver, concentration of	26–27
Sources of information	1–2
South Ibex stockwork, description of	12
Specularite, indications from	10
Stocks, mineralization around	8–10
Stockworks, structure and origin of	12–13
Stringers, indications from	22–23
Sulphide enrichment, results of	28
Sunday shaft, prospecting near	11
Tucson-Maid reverse fault, channels along and near	15–20
offsets by other faults, plan showing	16
relation of, to ore bodies in the Graham Park area, geologic sections showing	20
Tucson mine, enrichment in	27
Tucson shaft, cross section through	20
Veins, trend of	11
"Weber grits," ore in	10, 11
Weston fault, features of	22

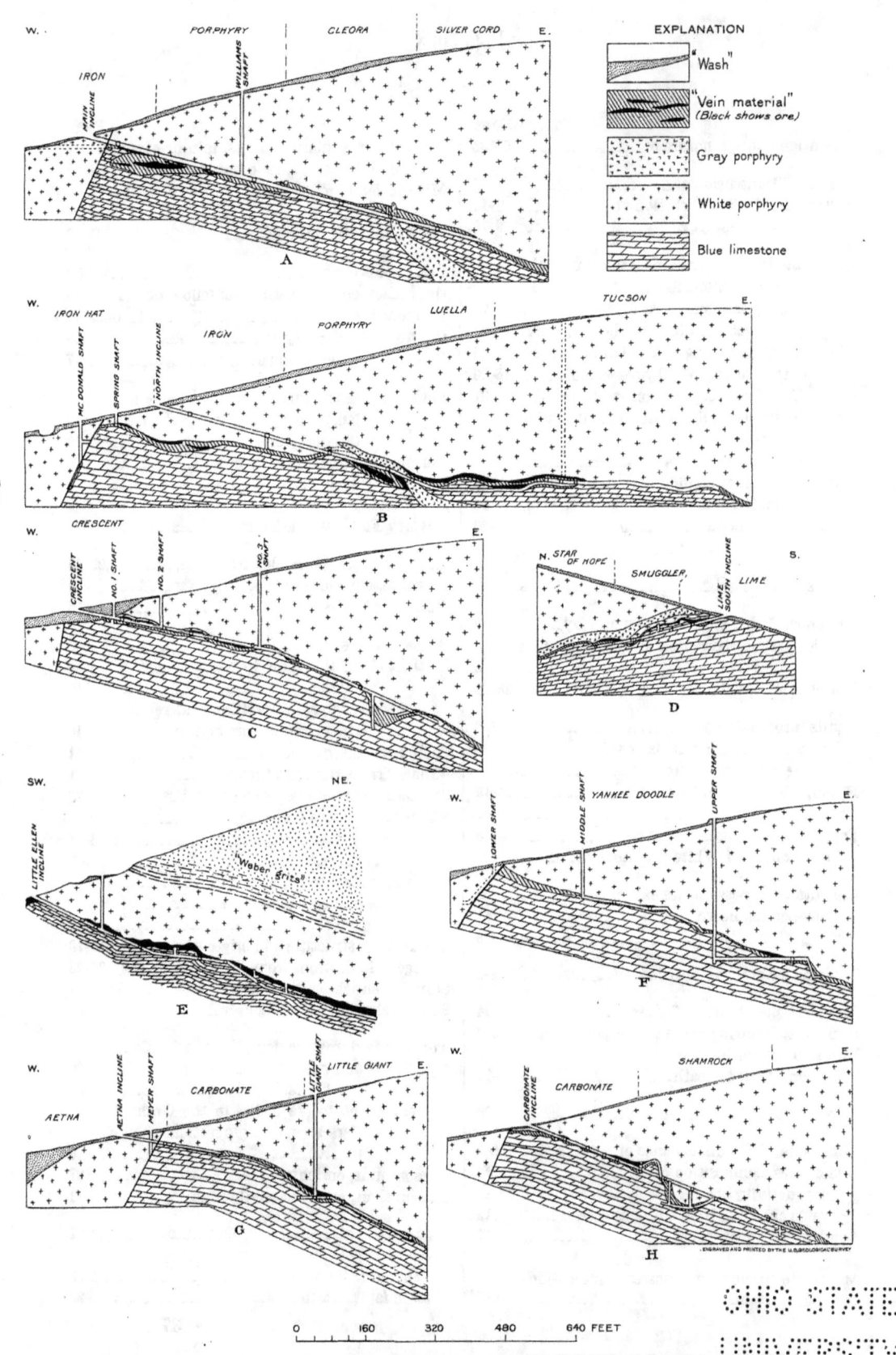

SECTIONS ALONG THE MOST IMPORTANT INCLINED SHAFTS IN THE LEADVILLE DISTRICT, COLO.

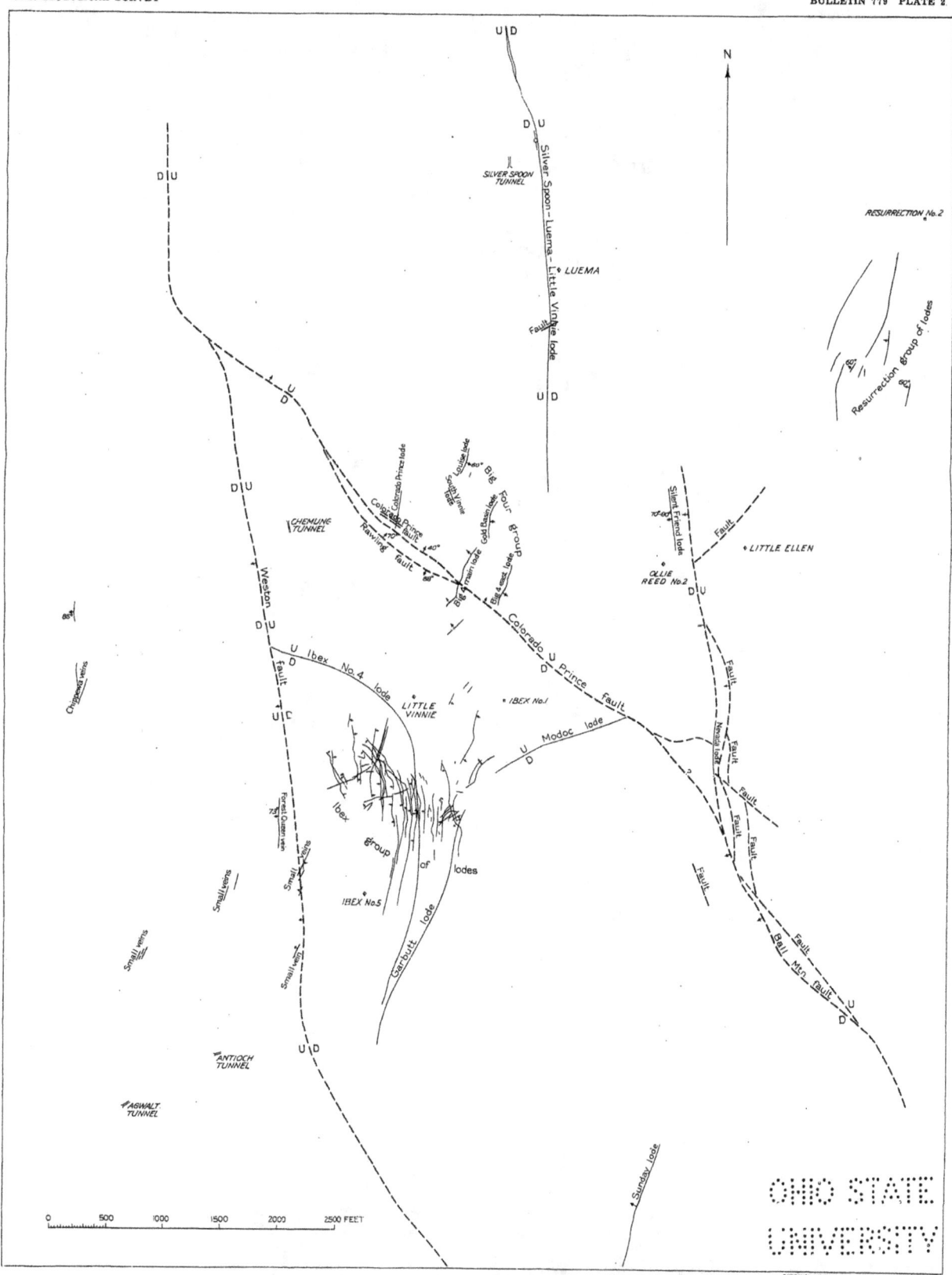

PLAN SHOWING THE MORE IMPORTANT LODES AND FAULTS IN THE EASTERN PART OF THE LEADVILLE DISTRICT, COLO.

COLUMNAR SECTIONS SHOWING POSITION AND RELATIVE SIZE OF BLANKET ORE BODIES IN DIFFERENT PARTS OF THE LEADVILLE DISTRICT, COLO.

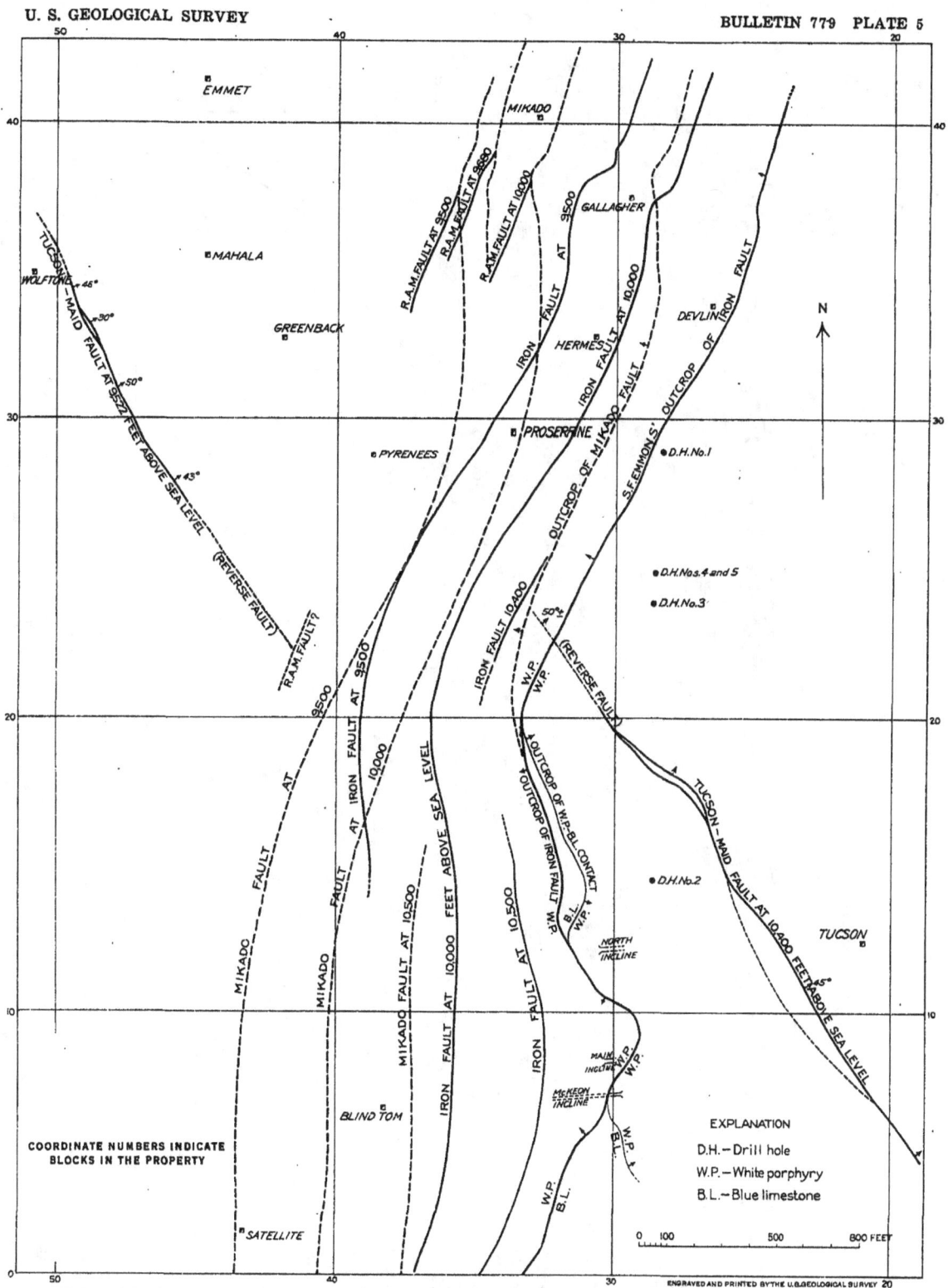

PLAN SHOWING TUCSON-MAID FAULT OFFSET BY THE IRON, MIKADO, AND R. A. M. FAULTS
by F. A. Aicher

CROSS SECTION THROUGH TUCSON SHAFT N. 63° E.
By F. A. Aicher

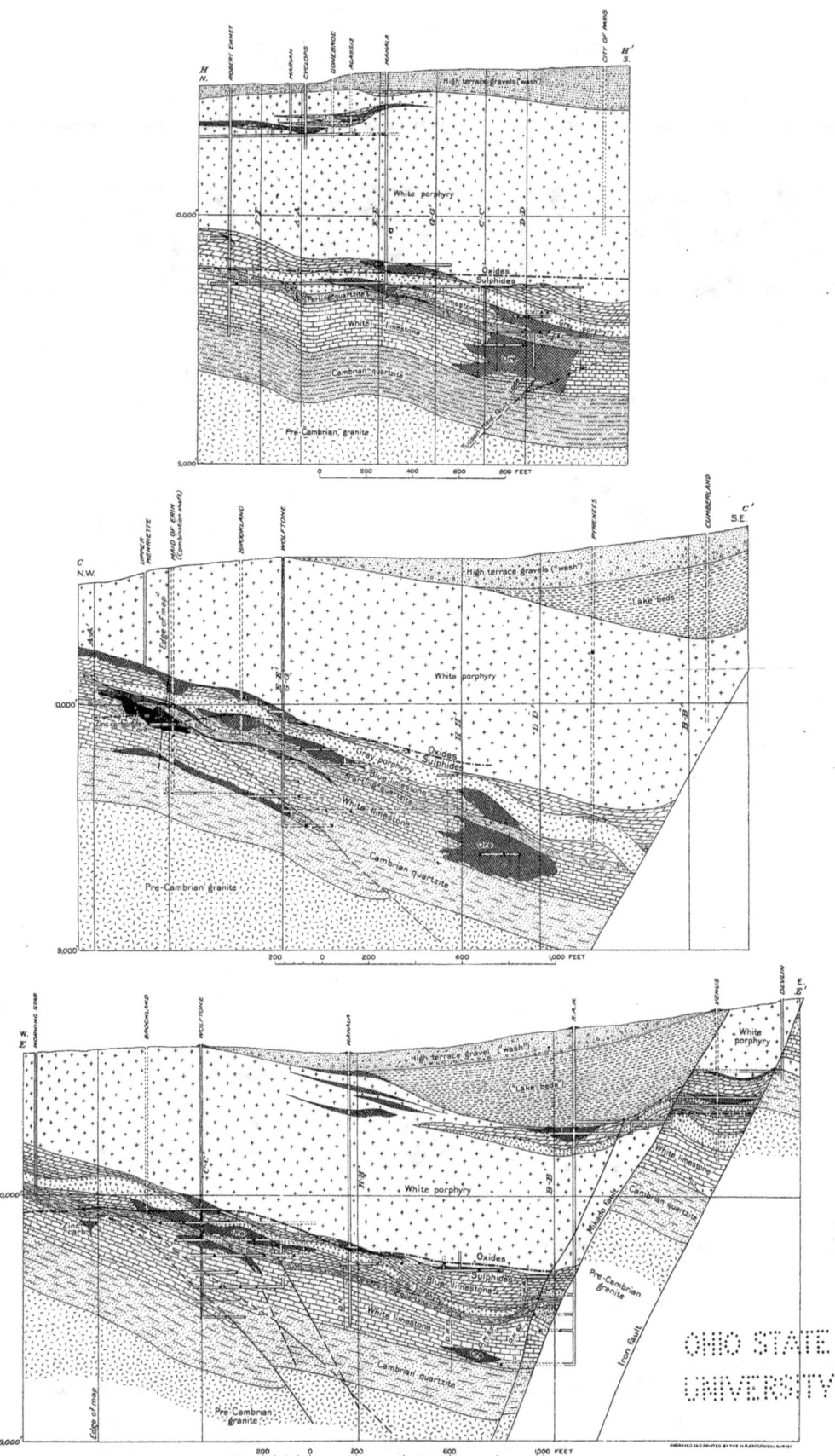

GEOLOGIC SECTIONS SHOWING GENERAL STRUCTURE AND RELATION OF TUCSON-MAID FAULT TO ORE BODIES IN THE GRAHAM PARK AREA

GOLD RUSH BOOKS

OREGON, USA

www.GoldMiningBooks.com

More Books On Mining

Visit: www.goldminingbooks.com to order your copies or ask your favorite book seller to offer them.

Mining Books by Kerby Jackson

Gold Dust: Stories From Oregon's Mining Years

Oregon mining historian and prospector, Kerby Jackson, brings you a treasure trove of seventeen stories on Southern Oregon's rich history of gold prospecting, the prospectors and their discoveries, and the breathtaking areas they settled in and made homes. 5" X 8", 98 ppgs. Retail Price: $11.99

The Golden Trail: More Stories From Oregon's Mining Years

In his follow-up to "Gold Dust: Stories of Oregon's Mining Years", this time around, Jackson brings us twelve tales from Oregon's Gold Rush, including the story about the first gold strike on Canyon Creek in Grant County, about the old timers who found gold by the pail full at the Victor Mine near Galice, how Iradel Bray discovered a rich ledge of gold on the Coquille River during the height of the Rogue River War, a tale of two elderly miners on the hunt for a lost mine in the Cascade Mountains, details about the discovery of the famous Armstrong Nugget and others. 5" X 8", 70 ppgs. Retail Price: $10.99

Oregon Mining Books

Geology and Mineral Resources of Josephine County, Oregon

Unavailable since the 1970's, this important publication was originally compiled by the Oregon Department of Geology and Mineral Industries and includes important details on the economic geology and mineral resources of this important mining area in South Western Oregon. Included are notes on the history, geology and development of important mines, as well as insights into the mining of gold, copper, nickel, limestone, chromium and other minerals found in large quantities in Josephine County, Oregon. 8.5" X 11", 54 ppgs. Retail Price: $9.99

Mines and Prospects of the Mount Reuben Mining District

Unavailable since 1947, this important publication was originally compiled by geologist Elton Youngberg of the Oregon Department of Geology and Mineral Industries and includes detailed descriptions, histories and the geology of the Mount Reuben Mining District in Josephine County, Oregon. Included are notes on the history, geology, development and assay statistics, as well as underground maps of all the major mines and prospects in the vicinity of this much neglected mining district. 8.5" X 11", 48 ppgs. Retail Price: $9.99

The Granite Mining District
Notes on the history, geology and development of important mines in the well known Granite Mining District which is located in Grant County, Oregon. Some of the mines discussed include the Ajax, Blue Ribbon, Buffalo, Continental, Cougar-Independence, Magnolia, New York, Standard and the Tillicum. Also included are many rare maps pertaining to the mines in the area. **8.5" X 11", 48 ppgs. Retail Price: $9.99**

Ore Deposits of the Takilma and Waldo Mining Districts of Josephine County, Oregon
The Waldo and Takilma mining districts are most notable for the fact that the earliest large scale mining of placer gold and copper in Oregon took place in these two areas. Included are details about some of the earliest large gold mines in the state such as the Llano de Oro, High Gravel, Cameron, Platerica, Deep Gravel and others, as well as copper mines such as the famous Queen of Bronze mine, the Waldo, Lily and Cowboy mines. This volume also includes six maps and 20 original illustrations. **8.5" X 11", 74 ppgs. Retail Price: $9.99**

Metal Mines of Douglas, Coos and Curry Counties, Oregon
Oregon mining historian Kerby Jackson introduces us to a classic work on Oregon's mining history in this important re-issue of Bulletin 14C Volume 1, otherwise known as the Douglas, Coos & Curry Counties, Oregon Metal Mines Handbook. Unavailable since 1940, this important publication was originally compiled by the Oregon Department of Geology and Mineral Industries includes detailed descriptions, histories and the geology of over 250 metallic mineral mines and prospects in this rugged area of South West Oregon. **8.5" X 11", 158 ppgs. Retail Price: $19.99**

Metal Mines of Jackson County, Oregon
Unavailable since 1943, this important publication was originally compiled by the Oregon Department of Geology and Mineral Industries includes detailed descriptions, histories and the geology of over 450 metallic mineral mines and prospects in Jackson County, Oregon. Included are such famous gold mining areas as Gold Hill, Jacksonville, Sterling and the Upper Applegate. **8.5" X 11", 220 ppgs. Retail Price: $24.99**

Metal Mines of Josephine County, Oregon
Oregon mining historian Kerby Jackson introduces us to a classic work on Oregon's mining history in this important re-issue of Bulletin 14C, otherwise known as the Josephine County, Oregon Metal Mines Handbook. Unavailable since 1952, this important publication was originally compiled by the Oregon Department of Geology and Mineral Industries includes detailed descriptions, histories and the geology of over 500 metallic mineral mines and prospects in Josephine County, Oregon. **8.5" X 11", 250 ppgs. Retail Price: $24.99**

Metal Mines of North East Oregon
Oregon mining historian Kerby Jackson introduces us to a classic work on Oregon's mining history in this important re-issue of Bulletin 14A and 14B, otherwise known as the North East Oregon Metal Mines Handbook. Unavailable since 1941, this important publication was originally compiled by the Oregon Department of Geology and Mineral Industries and includes detailed descriptions, histories and the geology of over 750 metallic mineral mines and prospects in North Eastern Oregon. **8.5" X 11", 310 ppgs. Retail Price: $29.99**

Metal Mines of North West Oregon

Oregon mining historian Kerby Jackson introduces us to a classic work on Oregon's mining history in this important re-issue of Bulletin 14D, otherwise known as the North West Oregon Metal Mines Handbook. Unavailable since 1951, this important publication was originally compiled by the Oregon Department of Geology and Mineral Industries and includes detailed descriptions, histories and the geology of over 250 metallic mineral mines and prospects in North Western Oregon. 8.5" X 11", 182 ppgs. Retail Price: $19.99

Mines and Prospects of Oregon

Mining historian Kerby Jackson introduces us to a classic mining work by the Oregon Bureau of Mines in this important re-issue of The Handbook of Mines and Prospects of Oregon. Unavailable since 1916, this publication includes important insights into hundreds of gold, silver, copper, coal, limestone and other mines that operated in the State of Oregon around the turn of the 19th Century. Included are not only geological details on early mines throughout Oregon, but also insights into their history, production, locations and in some cases, also included are rare maps of their underground workings. 8.5" X 11", 314 ppgs. Retail Price: $24.99

Lode Gold of the Klamath Mountains of Northern California and South West Oregon

(See California Mining Books)

Mineral Resources of South West Oregon

Unavailable since 1914, this publication includes important insights into dozens of mines that once operated in South West Oregon, including the famous gold fields of Josephine and Jackson Counties, as well as the Coal Mines of Coos County. Included are not only geological details on early mines throughout South West Oregon, but also insights into their history, production and locations. 8.5" X 11", 154 ppgs. Retail Price: $11.99

Chromite Mining in The Klamath Mountains of California and Oregon

(See California Mining Books)

Southern Oregon Mineral Wealth

Unavailable since 1904, this rare publication provides a unique snapshot into the mines that were operating in the area at the time. Included are not only geological details on early mines throughout South West Oregon, but also insights into their history, production and locations. Some of the mining areas include Grave Creek, Greenback, Wolf Creek, Jump Off Joe Creek, Granite Hill, Galice, Mount Reuben, Gold Hill, Galls Creek, Kane Creek, Sardine Creek, Birdseye Creek, Evans Creek, Foots Creek, Jacksonville, Ashland, the Applegate River, Waldo, Kerby and the Illinois River, Althouse and Sucker Creek, as well as insights into local copper mining and other topics. 8.5" X 11", 64 ppgs. Retail Price: $8.99

Geology and Ore Deposits of the Takilma and Waldo Mining Districts

Unavailable since the 1933, this publication was originally compiled by the United States Geological Survey and includes details on gold and copper mining in the Takilma and Waldo Districts of Josephine County, Oregon. The Waldo and Takilma mining districts are most notable for the fact that the earliest large scale mining of placer gold and copper in Oregon took place in these two areas. Included in this report are details about some of the earliest large gold mines in the state such as the Llano de Oro, High Gravel, Cameron, Platerica, Deep Gravel and others, as well as copper mines such as the famous Queen of Bronze mine, the Waldo, Lily and Cowboy mines. In addition to geological examinations, insights are also provided into the production, day to day operations and early histories of these mines, as well as calculations of known mineral reserves in the area. This volume also includes six maps and 20 original illustrations. **8.5" X 11", 74 ppgs. Retail Price: $9.99**

Gold Mines of Oregon

Oregon mining historian Kerby Jackson introduces us to a classic work on Oregon's mining history in this important re-issue of Bulletin 61, otherwise known as "Gold and Silver In Oregon". Unavailable since 1968, this important publication was originally compiled by geologists Howard C. Brooks and Len Ramp of the Oregon Department of Geology and Mineral Industries and includes detailed descriptions, histories and the geology of over 450 gold mines Oregon. Included are notes on the history, geology and gold production statistics of all the major mining areas in Oregon including the Klamath Mountains, the Blue Mountains and the North Cascades. While gold is where you find it, as every miner knows, the path to success is to prospect for gold where it was previously found. **8.5" X 11", 344 ppgs. Retail Price: $24.99**

Mines and Mineral Resources of Curry County Oregon

Originally published in 1916, this important publication on Oregon Mining has not been available for nearly a century. Included are rare insights into the history, production and locations of dozens of gold mines in Curry County, Oregon, as well as detailed information on important Oregon mining districts in that area such as those at Agness, Bald Face Creek, Mule Creek, Boulder Creek, China Diggings, Collier Creek, Elk River, Gold Beach, Rock Creek, Sixes River and elsewhere. Particular attention is especially paid to the famous beach gold deposits of this portion of the Oregon Coast. **8.5" X 11", 140 ppgs. Retail Price: $11.99**

Chromite Mining in South West Oregon

Originally published in 1961, this important publication on Oregon Mining has not been available for nearly a century. Included are rare insights into the history, production and locations of nearly 300 chromite mines in South Western Oregon. **8.5" X 11", 184 ppgs. Retail Price: $14.99**

Mineral Resources of Douglas County Oregon

Originally published in 1972, this important publication on Oregon Mining has not been available for nearly forty years. Included are rare insights into the geology, history, production and locations of numerous gold mines and other mining properties in Douglas County, Oregon. **8.5" X 11", 124 ppgs. Retail Price: $11.99**

Mineral Resources of Coos County Oregon

Originally published in 1972, this important publication on Oregon Mining has not been available for nearly forty years. Included are rare insights into the geology, history, production and locations of numerous gold mines and other mining properties in Coos County, Oregon. **8.5" X 11", 100 ppgs. Retail Price: $11.99**

Mineral Resources of Lane County Oregon

Originally published in 1938, this important publication on Oregon Mining has not been available for nearly seventy five years. Included are extremely rare insights into the geology and mines of Lane County, Oregon, in particular in the Bohemia, Blue River, Oakridge, Black Butte and Winberry Mining Districts. **8.5" X 11", 82 ppgs. Retail Price: $9.99**

Mineral Resources of the Upper Chetco River of Oregon: Including the Kalmiopsis Wilderness

Originally published in 1975, this important publication on Oregon Mining has not been available for nearly forty years. Withdrawn under the 1872 Mining Act since 1984, real insight into the minerals resources and mines of the Upper Chetco River has long been unavailable due to the remoteness of the area. Despite this, the decades of battle between property owners and environmental extremists over the last private mining inholding in the area has continued to pique the interest of those interested in mining and other forms of natural resource use. Gold mining began in the area in the 1850's and has a rich history in this geographic area, even if the facts surrounding it are little known. Included are twenty two rare photographs, as well as insights into the Becca and Morning Mine, the Emmly Mine (also known as Emily Camp), the Frazier Mine, the Golden Dream or Higgins Mine, Hustis Mine, Peck Mine and others. **8.5" X 11", 64 ppgs. Retail Price: $8.99**

Gold Dredging in Oregon

Originally published in 1939, this important publication on Oregon Mining has not been available for nearly seventy five years. Included are extremely rare insights into the history and day to day operations of the dragline and bucketline gold dredges that once worked the placer gold fields of South West and North East Oregon in decades gone by. Also included are details into the areas that were worked by gold dredges in Josephine, Jackson, Baker and Grant counties, as well as the economic factors that impacted this mining method. This volume also offers a unique look into the values of river bottom land in relation to both farming and mining, in how farm lands were mined, re-soiled and reclamated after the dredges worked them. Featured are hard to find maps of the gold dredge fields, as well as rare photographs from a bygone era. **8.5" X 11", 86 ppgs. Retail Price: $8.99**

Quick Silver Mining in Oregon

Originally published in 1963, this important publication on Oregon Mining has not been available for over fifty years. This publication includes details into the history and production of Elemental Mercury or Quicksilver in the State of Oregon. **8.5" X 11", 238 ppgs. Retail Price: $15.99**

Mines of the Greenhorn Mining District of Grant County Oregon

Originally published in 1948, this important publication on Oregon Mining has not been available for over sixty five years. In this publication are rare insights into the mines of the famous Greenhorn Mining District of Grant County, Oregon, especially the famous Morning Mine. Also included are details on the Tempest, Tiger, Bi-Metallic, Windsor, Psyche, Big Johnny, Snow Creek, Banzette and Paramount Mines, as well as prospects in the vicinities in the famous mining areas of Mormon Basin, Vinegar Basin and Desolation Creek. Included are hard to find mine maps and dozens of rare photographs from the bygone era of Grant County's rich mining history. **8.5" X 11", 72 ppgs. Retail Price: $9.99**

Geology of the Wallowa Mountains of Oregon: Part I (Volume 1)

Originally published in 1938, this important publication on Oregon Mining has not been available for nearly seventy five years. Included are details on the geology of this unique portion of North Eastern Oregon. This is the first part of a two book series on the area. Accompanying the text are rare photographs and historic maps. 8.5" X 11", 92 ppgs. Retail Price: $9.99

Geology of the Wallowa Mountains of Oregon: Part II (Volume 2)

Originally published in 1938, this important publication on Oregon Mining has not been available for nearly seventy five years. Included are details on the geology of this unique portion of North Eastern Oregon. This is the first part of a two book series on the area. Accompanying the text are rare photographs and historic maps. 8.5" X 11", 94 ppgs. Retail Price: $9.99

Field Identification of Minerals For Oregon Prospectors

Originally published in 1940, this important publication on Oregon Mining has not been available for nearly seventy five years. Included in this volume is an easy system for testing and identifying a wide range of minerals that might be found by prospectors, geologists and rockhounds in the State of Oregon, as well as in other locales. Topics include how to put together your own field testing kit and how to conduct rudimentary tests in the field. This volume is written in a clear and concise way to make it useful even for beginners. 8.5" X 11", 158 ppgs. Retail Price: $14.99

Idaho Mining Books

Gold in Idaho

Unavailable since the 1940's, this publication was originally compiled by the Idaho Bureau of Mines and includes details on gold mining in Idaho. Included is not only raw data on gold production in Idaho, but also valuable insight into where gold may be found in Idaho, as well as practical information on the gold bearing rocks and other geological features that will assist those looking for placer and lode gold in the State of Idaho. This volume also includes thirteen gold maps that greatly enhance the practical usability of the information contained in this small book detailing where to find gold in Idaho. 8.5" X 11", 72 ppgs. Retail Price: $9.99

Geology of the Couer D'Alene Mining District of Idaho

Unavailable since 1961, this publication was originally compiled by the Idaho Bureau of Mines and Geology and includes details on the mining of gold, silver and other minerals in the famous Coeur D'Alene Mining District in Northern Idaho. Included are details on the early history of the Coeur D'Alene Mining District, local tectonic settings, ore deposit features, information on the mineral belts of the Osburn Fault, as well as detailed information on the famous Bunker Hill Mine, the Dayrock Mine, Galena Mine, Lucky Friday Mine and the infamous Sunshine Mine. This volume also includes sixteen hard to find maps. 8.5" X 11", 70 ppgs. Retail Price: $9.99

The Gold Camps and Silver Cities of Idaho

Originally published in 1963, this important publication on Idaho Mining has not been available for nearly fifty years. Included are rare insights into the history of Idaho's Gold Rush, as well as the mad craze for silver in the Idaho Panhandle. Documented in fine detail are the early mining excitements at Boise Basin, at South Boise, in the Owyhees, at Deadwood, Long Valley, Stanley Basin and Robinson Bar, at Atlanta, on the famous Boise River, Volcano, Little Smokey, Banner, Boise Ridge, Hailey, Leesburg, Lemhi, Pearl, at South Mountain, Shoup and Ulysses, Yellow Jacket and Loon Creek. The story follows with the appearance of Chinese miners at the new mining camps on the Snake River, Black Pine, Yankee Fork, Bay Horse, Clayton, Heath, Seven Devils, Gibbonsville, Vienna and Sawtooth City. Also included are special sections on the Idaho Lead and Silver mines of the late 1800's, as well as the mining discoveries of the early 1900's that paved the way for Idaho's modern mining and mineral industry. Lavishly illustrated with rare historic photos, this volume provides a one of a kind documentary into Idaho's mining history that is sure to be enjoyed by not only modern miners and prospectors who still scour the hills in search of nature's treasures, but also those enjoy history and tromping through overgrown ghost towns and long abandoned mining camps. **8.5" X 11", 186 ppgs. Retail Price: $14.99**

Utah Mining Books

Fluorite in Utah

Unavailable since 1954, this publication was originally compiled by the USGS, State of Utah and U.S. Atomic Energy Commission and details the mining of fluorspar, also known as fluorite in the State of Utah. Included are details on the geology and history of fluorspar (fluorite) mining in Utah, including details on where this unique gem mineral may be found in the State of Utah. **8.5" X 11", 60 ppgs. Retail Price: $8.99**

California Mining Books

The Tertiary Gravels of the Sierra Nevada of California

Mining historian Kerby Jackson introduces us to a classic mining work by Waldemar Lindgren in this important re-issue of The Tertiary Gravels of the Sierra Nevada of California. Unavailable since 1911, this publication includes details on the gold bearing ancient river channels of the famous Sierra Nevada region of California. **8.5" X 11", 282 ppgs. Retail Price: $19.99**

The Mother Lode Mining Region of California

Unavailable since 1900, this publication includes details on the gold mines of California's famous Mother Lode gold mining area. Included are details on the geology, history and important gold mines of the region, as well as insights into historic mining methods, mine timbering, mining machinery, mining bell signals and other details on how these mines operated. Also included are insights into the gold mines of the California Mother Lode that were in operation during the first sixty years of California's mining history. **8.5" X 11", 176 ppgs. Retail Price: $14.99**

Lode Gold of the Klamath Mountains of Northern California and South West Oregon

Unavailable since 1971, this publication was originally compiled by Preston E. Hotz and includes details on the lode mining districts of Oregon and California's Klamath Mountains. Included are details on the geology, history and important lode mines of the French Gulch, Deadwood, Whiskeytown, Shasta, Redding, Muletown, South Fork, Old Diggings, Dog Creek (Delta), Bully Choop (Indian Creek), Harrison Gulch, Hayfork, Minersville, Trinity Center, Canyon Creek, East Fork, New River, Denny, Liberty (Black Bear), Cecilville, Callahan, Yreka, Fort Jones and Happy Camp mining districts in California, as well as the Ashland, Rogue River, Applegate, Illinois River, Takilma, Greenback, Galice, Silver Peak, Myrtle Creek and Mule Creek districts of South Western Oregon. Also included are insights into the mineralization and other characteristics of this important mining region. **8.5" X 11", 100 ppgs. Retail Price: $10.99**

Mines and Mineral Resources of Shasta County, Siskiyou County, Trinity County, California

Unavailable since 1915, this publication was originally compiled by the California State Mining Bureau and includes details on the gold mines of this area of Northern California. Also included are insights into the mineralization and other characteristics of this important mining region, as well as the location of historic gold mines. **8.5" X 11", 204 ppgs. Retail Price: $19.99**

Geology of the Yreka Quadrangle, Siskiyou County, California

Unavailable since 1977, this publication was originally compiled by Preston E. Hotz and includes details on the geology of the Yreka Quadrangle of Siskiyou County, California. Also included are insights into the mineralization and other characteristics of this important mining region. **8.5" X 11", 78 ppgs. Retail Price: $7.99**

Mines of San Diego and Imperial Counties, California

Originally published in 1914, this important publication on California Mining has not been available for a century. This publication includes important information on the early gold mines of San Diego and Imperial County, which were some of the first gold fields mined in California by early Spanish and Mexican miners before the 49ers came on the scene. Included are not only details on early mining methods in the area, production statistics and geological information, but also the location of the early gold mines that helped make California "The Golden State". Also included are details on the mining of other minerals such as silver, lead, zinc, manganese, tungsten, vanadium, asbestos, barite, borax, cement, clay, dolomite, fluospar, gem stones, graphite, marble, salines, petroleum, stronium, talc and others. **8.5" X 11", 116 ppgs. Retail Price: $12.99**

Mines of Sierra County, California

Unavailable since 1920, this publication was originally compiled by the California State Mining Bureau and includes details on the gold mines of Sierra County, California. Also included are insights into the mineralization and other characteristics of this important mining region, as well as the location of historic gold mines. **8.5" X 11", 156 ppgs. Retail Price: $19.99**

Mines of Plumas County, California

Unavailable since 1918, this publication was originally compiled by the California State Mining Bureau and includes details on the gold mines of Plumas County, California. Also included are insights into the mineralization and other characteristics of this important mining region, as well as the location of historic gold mines. **8.5" X 11", 200 ppgs. Retail Price: $19.99**

Mines of El Dorado, Placer, Sacramento and Yuba Counties, California

Originally published in 1917, this important publication on California Mining has not been available for nearly a century. This publication includes important information on the early gold mines of El Dorado County, Placer County, Sacramento County and Yuba County, which were some of the first gold fields mined by the Forty-Niners during the California Gold Rush. Included are not only details on early mining methods in the area, production statistics and geological information, but also the location of the early gold mines that helped make California "The Golden State". Also included are insights into the early mining of chrome, copper and other minerals in this important mining area. **8.5" X 11", 204 ppgs. Retail Price: $19.99**

Mines of Los Angeles, Orange and Riverside Counties, California

Originally published in 1917, this important publication on California Mining has not been available for nearly a century. This publication includes important information on the early gold mines of Los Angeles County, Orange County and Riverside County, which were some of the first gold fields mined in California by early Spanish and Mexican miners before the 49ers came on the scene. Included are not only details on early mining methods in the area, production statistics and geological information, but also the location of the early gold mines that helped make California "The Golden State". **8.5" X 11", 146 ppgs. Retail Price: $12.99**

Mines of San Bernadino and Tulare Counties, California

Originally published in 1917, this important publication on California Mining has not been available for nearly a century. This publication includes important information on the early gold mines of San Bernadino and Tulare County, which were some of the first gold fields mined in California by early Spanish and Mexican miners before the 49ers came on the scene. Included are not only details on early mining methods in the area, production statistics and geological information, but also the location of the early gold mines that helped make California "The Golden State". Also included are details on the mining of other minerals such as copper, iron, lead, zinc, manganese, tungsten, vanadium, asbestos, barite, borax, cement, clay, dolomite, fluospar, gem stones, graphite, marble, salines, petroleum, stronium, talc and others. **8.5" X 11", 200 ppgs. Retail Price: $19.99**

Chromite Mining in The Klamath Mountains of California and Oregon

Unavailable since 1919, this publication was originally compiled by J.S. Diller of the United States Department of Geological Survey and includes details on the chromite mines of this area of Northern California and Southern Oregon. Also included are insights into the mineralization and other characteristics of this important mining region, as well as the location of historic mines. Also included are insights into chromite mining in Eastern Oregon and Montana. **8.5" X 11", 98 ppgs. Retail Price: $9.99**

Mines and Mining in Amador, Calaveras and Tuolumne Counties, California

Unavailable since 1915, this publication was originally compiled by William Tucker and includes details on the mines and mineral resources of this important California mining area. Included are details on the geology, history and important gold mines of the region, as well as insights into other local mineral resources such as asbestos, clay, copper, talc, limestone and others. Also included are insights into the mineralization and other characteristics of this important portion of California's Mother Lode mining region. **8.5" X 11", 198 ppgs. Retail Price: $14.99**

The Cerro Gordo Mining District of Inyo County California

Unavailable since 1963, this publication was originally compiled by the United States Department of Interior. Included are insights into the mineralization and other characteristics of this important mining region of Southern California. Topics include the mining of gold and silver in this important mining district in Inyo County, California, including details on the history, production and locations of the Cerro Gordo Mine, the Morning Star Mine, Estelle Tunnel, Charles Lease Tunnel, Ignacio, Hart, Crosscut Tunnel, Sunset, Upper Newtown, Newtown, Ella, Perseverance, Newsboy, Belmont and other silver and gold mines in the Cerro Gordo Mining District. This volume also includes important insights into the fossil record, geologic formations, faults and other aspects of economic geology in this California mining district. **8.5" X 11", 104 ppgs. Retail Price: $10.99**

Mining in Butte, Lassen, Modoc, Sutter and Tehama Counties of California

Unavailable since 1917, this publication was originally compiled by the United States Department of Interior. Included are insights into the mineralization and other characteristics of this important mining region of California. Topics include the mining of asbestos, chromite, gold, diamonds and manganese in Butte County, the mining of gold and copper in the Hayden Hill and Diamond Mountain mining districts of Lassen County, the mining of coal, salt, copper and gold in the High Grade and Winters mining districts of Modoc County, gold mining in Sutter County and the mining of gold, chromite, manganese and copper in Tehama County. This volume also includes the production records and locations of numerous mines in this important mining region. **8.5" X 11", 114 ppgs. Retail Price: $11.99**

Alaska Mining Books

Ore Deposits of the Willow Creek Mining District, Alaska

Unavailable since 1954, this hard to find publication includes valuable insights into the Willow Creek Mining District near Hatcher Pass in Alaska. The publication includes insights into the history, geology and locations of the well known mines in the area, including the Gold Cord, Independence, Fern, Mabel, Lonesome, Snowbird, Schroff-O'Neil, High Grade, Marion Twin, Thorpe, Webfoot, Kelly-Willow, Lane, Holland and others. **8.5" X 11", 96 ppgs. Retail Price: $9.99**

Arizona Mining Books

Mines and Mining in Northern Yuma County Arizona

Originally published in 1911, this important publication on Arizona Mining has not been available for over a hundred years. Included are rare insights into the gold, silver, copper and quicksilver mines of Yuma County, Arizona together with hard to find maps and photographs. Some of the mines and mining districts featured include the Planet Copper Mine, Mineral Hill, the Clara Consolidated Mine, Viati Mine, Copper Basin prospect, Bowman Mine, Quartz King, Billy Mack, Carnation, the Wardwell and Osbourne, Valensuella Copper, the Mariquita, Colonial Mine, the French American, the New York-Plomosa, Guadalupe, Lead Camp, Mudersbach Copper Camp, Yellow Bird, the Arizona Northern (Salome Strike), Bonanza (Harqua Hala), Golden Eagle, Hercules, Socorro and others. **8.5" X 11", 144 ppgs. Retail Price: $11.99**

The Aravaipa and Stanley Mining Districts of Graham County Arizona

Originally published in 1925, this important publication on Arizona Mining has not been available for nearly ninety years. Included are rare insights into the gold and silver mines of these two important mining districts, together with hard to find maps. **8.5" X 11", 140 ppgs. Retail Price: $11.99**

Gold in the Gold Basin and Lost Basin Mining Districts of Mohave County, Arizona

This volume contains rare insights into the geology and gold mineralization of the Gold Basin and Lost Basin Mining Districts of Mohave County, Arizona that will be of benefit to miners and prospectors. Also included is a significant body of information on the gold mines and prospects of this portion of Arizona. This volume is lavishly illustrated with rare photos and mining maps. **8.5" X 11", 188 ppgs. Retail Price: $19.99**

Mines of the Jerome and Bradshaw Mountains of Arizona

This important publication on Arizona Mining has not been available for ninety years. This volume contains rare insights into the geology and ore deposits of the Jerome and Bradshaw Mountains of Arizona that will be of benefit to miners and prospectors who work those areas. Included is a significant body of information on the mines and prospects of the Verde, Black Hills, Cherry Creek, Prescott, Walker, Groom Creek, Hassayampa, Bigbug, Turkey Creek, Agua Fria, Black Canyon, Peck, Tiger, Pine Grove, Bradshaw, Tintop, Humbug and Castle Creek Mining Districts. This volume is lavishly illustrated with rare photos and mining maps. **8.5" X 11", 218 ppgs. Retail Price: $19.99**

The Ajo Mining District of Pima County Arizona

This important publication on Arizona Mining has not been available for nearly seventy years. This volume contains rare insights into the geology and mineralization of the Ajo Mining District in Pima County, Arizona and in particular the famous New Cornelia Mine.   **8.5" X 11", 126 ppgs. Retail Price: $11.99**

Mining in the Santa Rita and Patagonia Mountains of Arizona

Originally published in 1915, this important publication on Arizona Mining has not been available for nearly a century. Included are rare insights into hundreds of gold, silver, copper and other mines in this famous Arizona mining area. Details include the locations, geology, history, production and other facts of the mines of this region. **8.5" X 11", 394 ppgs. Retail Price: $24.99**

Montana Mining Books

A History of Butte Montana: The World's Greatest Mining Camp

First published in 1900 by H.C. Freeman, this important publication sheds a bright light on one of the most important mining areas in the history of The West. Together with his insights, as well as rare photographs of the periods, Harry Freeman describes Butte and its vicinity from its early beginnings, right up to its flush years when copper flowed from its mines like a river. At the time of publication, Butte, Montana was known worldwide as "The Richest Mining Spot On Earth" and produced not only vast amounts of copper, but also silver, gold and other metals from its mines. Freeman illustrates, with great detail, the most important mines in the vicinity of Butte, providing rare details on their owners, their history and most importantly, how the mines operated and how their treasures were extracted. Of particular interest are the dozens of rare photographs that depict mines such as the famous Anaconda, the Silver Bow, the Smoke House, Moose, Paulin, Buffalo, Little Minah, the Mountain Consolidated, West Greyrock, Cora, the Green Mountain, Diamond, Bell, Parnell, the Neversweat, Nipper, Original and many others.   **8.5" X 11", 142 ppgs. Retail Price: $12.99**

The Butte Mining District of Montana

This important publication on Montana Mining has not been available for over a century. Included are rare insights into the gold, copper and silver mines of Butte, Montana together with hard to find maps and photographs. Some of the topics include the early history of gold, silver and copper mining in the Butte area, insight into the geology of its mining areas, the local distribution of gold, silver and copper ores, as well their composition and how to identify them. Also included are detailed facts about the mines in the Butte Mining District, including the famous Anaconda Mine, Gagnon, Parrot, Blue Vein, Moscow, Poulin, Stella, Buffalo, Green Mountain, Wake Up Jim, the Diamond-Bell Group, Mountain Consolidated, East Greyrock, West Greyrock, Snowball, Corra, Speculator, Adirondack, Miners Union, the Jessie-Edith May Group, Otisco, Iduna, Colorado, Lizzie, Cambers, Anderson, Hesperus, Preferencia and dozens of others. **8.5" X 11", 298 ppgs. Retail Price: $24.99**

Mines of the Helena Mining Region of Montana

This important publication on Montana Mining has not been available for over a century. Included are rare insights into the gold, copper and silver mines of the vicinity of Helena, Montana, including the Marysville Mining District, Elliston Mining District, Rimini Mining District, Helena Mining District, Clancy Mining District, Wickes Mining District, Boulder and Basin Mining Districts and the Elkhorn Mining District. Some of the topics include the early history of gold, silver and copper mining in the Helena area, insight into the geology of its mining areas, the local distribution of gold, silver and copper ores, as well their composition and how to identify them. Also included are detailed facts, history, geology and locations of over one hundred gold, silver and copper mines in the area . **8.5" X 11", 162 ppgs, Retail Price: $14.99**

Mines and Geology of the Garnet Range of Montana

This important publication on Montana Mining has not been available for over a century. Included are rare insights into the gold, copper and silver mines of the vicinity of this important mining area of Montana. Some of the topics include the early history of gold, silver and copper mining in the Garnet Mountains, insight into the geology of its mining areas, the local distribution of gold, silver and copper ores, as well their composition and how to identify them. Also included are detailed facts, history, geology and locations of numerous gold, silver and copper mines in the area . **8.5" X 11", 100 ppgs, Retail Price: $11.99**

Mines and Geology of the Philipsburg Quadrangle of Montana

This important publication on Montana Mining has not been available for over a century. Included are rare insights into the gold, copper and silver mines of the vicinity of this important mining area of Montana. Some of the topics include the early history of gold, silver and copper mining in the Philipsburg Quadrangle, insight into the geology of its mining areas, the local distribution of gold, silver and copper ores, as well their composition and how to identify them. Also included are detailed facts, history, geology and locations of over one hundred gold, silver and copper mines in the area **8.5" X 11", 290 ppgs, Retail Price: $24.99**

Geology of the Marysville Mining District of Montana

Included are rare insights into the mining geology of the Marysville Mining District. Some of the topics include the early history of gold, silver and copper mining in the area, insight into the geology of its mining areas, the local distribution of gold, silver and copper ores, as well their composition and how to identify them. Also included are detailed facts, history, geology and locations of gold, silver and copper mines in the area **8.5" X 11", 198 ppgs, Retail Price: $19.99**

More Mining Books

Prospecting and Developing A Small Mine

Topics covered include the classification of varying ores, how to take a proper ore sample, the proper reduction of ore samples, alluvial sampling, how to understand geology as it is applied to prospecting and mining, prospecting procedures, methods of ore treatment, the application of drilling and blasting in a small mine and other topics that the small scale miner will find of benefit. **8.5" X 11", 112 ppgs, Retail Price: $11.99**

Timbering For Small Underground Mines

Topics covered include the selection of caps and posts, the treatment of mine timbers, how to install mine timbers, repairing damaged timbers, use of drift supports, headboards, squeeze sets, ore chute construction, mine cribbing, square set timbering methods, the use of steel and concrete sets and other topics that the small underground miner will find of benefit. This volume also includes twenty eight illustrations depicting the proper construction of mine timbering and support systems that greatly enhance the practical usability of the information contained in this small book. **8.5" X 11", 88 ppgs. Retail Price: $10.99**

Timbering and Mining

A classic mining publication on Hard Rock Mining by W.H. Storms. Unavailable since 1909, this rare publication provides an in depth look at American methods of underground mine timbering and mining methods. Topics include the selection and preservation of mine timbers, drifting and drift sets, driving in running ground, structural steel in mine workings, timbering drifts in gravel mines, timbering methods for driving shafts, positioning drill holes in shafts, timbering stations at shafts, drainage, mining large ore bodies by means of open cuts or by the "Glory Hole" system, stoping out ore in flat or low lying veins, use of the "Caving System", stoping in swelling ground, how to stope out large ore bodies, Square Set timbering on the Comstock and its modifications by California miners, the construction of ore chutes, stoping ore bodies by use of the "Block System", how to work dangerous ground, information on the "Delprat System" of stoping without mine timbers, construction and use of headframes and much more. This volume provides a reference into not only practical methods of mining and timbering that may be employed in narrow vein mining by small miners today, but also rare insights into how mines were being worked at the turn of the 19th Century. **8.5" X 11", 288 ppgs. Retail Price: $24.99**

A Study of Ore Deposits For The Practical Miner

Mining historian Kerby Jackson introduces us to a classic mining publication on ore deposits by J.P. Wallace. First published in 1908, it has been unavailable for over a century. Included are important insights into the properties of minerals and their identification, on the occurrence and origin of gold, on gold alloys, insights into gold bearing sulfides such as pyrites and arsenopyrites, on gold bearing vanadium, gold and silver tellurides, lead and mercury tellurides, on silver ores, platinum and iridium, mercury ores, copper ores, lead ores, zinc ores, iron ores, chromium ores, manganese ores, nickel ores, tin ores, tungsten ores and others. Also included are facts regarding rock forming minerals, their composition and occurrences, on igneous, sedimentary, metamorphic and intrusive rocks, as well as how they are geologically disturbed by dikes, flows and faults, as well as the effects of these geologic actions and why they are important to the miner. Written specifically with the common miner and prospector in mind, the book will help to unlock the earth's hidden wealth for you and is written in a simple and concise language that anyone can understand. **8.5" X 11", 366 ppgs. Retail Price: $24.99**

Mine Drainage

Unavailable since 1896, this rare publication provides an in depth look at American methods of underground mine drainage and mining pump systems. This volume provides a reference into not only practical methods of mining drainage that may be employed in narrow vein mining by small miners today, but also rare insights into how mines were being worked at the turn of the 19th Century. **8.5" X 11", 218 ppgs. Retail Price: $24.99**

Fire Assaying Gold, Silver and Lead Ores

Unavailable since 1907, this important publication was originally published by the Mining and Scientific Press and was designed to introduce miners and prospectors of gold, silver and lead to the art of fire assaying. Topics include the fire assaying of ores and products containing gold, silver and lead; the sampling and preparation of ore for an assay; care of the assay office, assay furnaces; crucibles and scorifiers; assay balances; metallic ores; scorification assays; cupelling; parting' crucible assays, the roasting of ores and more. This classic provides a time honored method of assaying put forward in a clear, concise and easy to understand language that will make it a benefit to even beginners. **8.5" X 11", 96 ppgs. Retail Price: $11.99**